DNA
NANOTECHNOLOGY
FOR BIOANALYSIS

From Hybrid DNA Nanostructures to Functional Devices

DNA
NANOTECHNOLOGY
FOR BIOANALYSIS

From Hybrid DNA Nanostructures to Functional Devices

Editors

Giuseppe Arrabito
University of Palermo, Italy

Liqian Wang
Shanghai Institute of Applied Physics,
Chinese Academy of Sciences, China

World Scientific

NEW JERSEY · LONDON · SINGAPORE · BEIJING · SHANGHAI · HONG KONG · TAIPEI · CHENNAI · TOKYO

Published by

World Scientific Publishing Europe Ltd.
57 Shelton Street, Covent Garden, London WC2H 9HE
Head office: 5 Toh Tuck Link, Singapore 596224
USA office: 27 Warren Street, Suite 401-402, Hackensack, NJ 07601

Library of Congress Cataloging-in-Publication Data
Names: Arrabito, Giuseppe, editor. | Wang, Liqian, editor.
Title: DNA nanotechnology for bioanalysis : from hybrid DNA nanostructures to
    functional devices / edited by Giuseppe Arrabito (University of Palermo, Italy),
    Liqian Wang (Chinese Academy of Sciences, China).
Description: New Jersey : World Scientific, [2017] | Includes bibliographical references.
Identifiers: LCCN 2017013235 | ISBN 9781786343796 (hc : alk. paper)
Subjects: LCSH: DNA--Biotechnology. | Nanobiotechnology.
Classification: LCC QP624 .D165 2017 | DDC 572.8/6--dc23
LC record available at https://lccn.loc.gov/2017013235

British Library Cataloguing-in-Publication Data
A catalogue record for this book is available from the British Library.

Desk Editor: Kalpana Bharanikumar

Typeset by Stallion Press
Email: enquiries@stallionpress.com

Dedication

Giuseppe Arrabito would like to dedicate this book to all the colleagues of the research group of Prof. Bruno Pignataro for their support during the writing phase. Giuseppe Arrabito would like to thank to Prof. Weihua Han (Lanzou University, China) for useful discussion.

Giuseppe Arrabito would finally like to dedicate this book to his parents, for their unconditional love.

Foreword

Deoxyribose Nucleic Acid (DNA) is the molecule that contains the life code as a "blue print" in all living systems. DNA is a perfect material for building nanostructures because of the special Watson–Crick pairing rule involving four fundamental bases (A, T, G, C). Each base can be considered as a basic brick for designing nanostructures. The concept is very much like a LEGO methodology of constructing a structure from basic bricks. Even if the basic elements are restricted to four, we can program an infinite number of different sequences of such elements and therefore, using the pairing rule, create an infinite number of structures.

It has been over 30 years since the pioneering work by Professor Nadrian Seeman of New York University in the 1980s. We have seen a tremendous change in this field, DNA nanotechnology is not just beautiful nanoscaled structures, it holds great promise in life science applications.

This book has been written by the researchers in different groups in Europe and China who work in the field. The presentation is lucid and simple; it is easily understandable by beginners and undergraduate students. This book provides both a solid background introduction and most advanced applications. I trust that students and researchers who are not in the field will find this book a useful introduction to DNA nanotechnology for bioanalysis.

Prof. Chunhai Fan
Laboratory of Physical Biology, Shanghai Institute of Applied Physics, Chinese Academy of Science, Shanghai 201800, China

Preface

Writing a book about the latest advances in DNA nanotechnology is definitely a very difficult task, since this is one of the most rapidly growing scientific topics and it is destined to produce a tremendous amount of new fundamental concepts in the world of Nanoscience and new class of molecular devices allowing for advances in Chemistry, Biology and Medicine. One has to consider that 542 papers about G quadruplex and 1564 papers (272 according to web of science) about DNA origami have been published in 2016, according to Scopus database. This means that the idea to provide a comprehensive view of the field is challenging.

The idea that originated the book came up whilst the editors, G. Arrabito and L. Wang, published a Critical Review on The Analyst which, for the first time, reported on applications of DNA nanotechnology in the bioanalytical sciences. Following up that work, they had the opportunity to cover this exciting field under another perspective — with the hope to reach a different readership, especially students and researchers who are not specifically expert in the DNA nanotechnology and its applications.

We here tried to explore the general aspects of functional DNA nanostructures such as DNA tiles, DNA origami, DNA quadruplexes, DNA-based bio- and inorganic covalent immobilization for biosensing applications and analytical sciences. We included the state of the art of important methods used for the mild, biocompatible and programmable fabrication of several different types of DNA nanostructures and hybrid nanostructures bearing proteins, molecules or inorganic moieties covalently binded onto DNA via self-assemblage processes. All the fabricated structures here reported are discussed according to the attributes that are of specific interest for applications in fields like biotechnology, biosensing, electrochemistry, molecular mimicking, diagnostic, biomedicine and analytical biochemistry.

The book can be considered as a clear and concise introduction to the field and is specifically focused on the new applications of DNA nanotechnology in sensors and Chemistry, especially in the field of Analytical Chemistry in which scientists are just starting to think about usability of DNA nanostructures. In fact, most of the papers published so far are in the material science field (750), whereas a fewer amount (30) is in the Analytical Chemistry category.

A part from providing an introduction into the exciting field of DNA nanosciences, the main aim of the book is the one to define some points of future development in the field, mainly — we concluded — driven by the need to make such technology more useful and cost-effective by solving issues like the high cost of oligos synthesis, better means of detection and efficient integration into output devices in order to fully show its potentiality in real world applications. We believe that DNA nanotechnology is still not at his mature age and more breakthroughs still have to come in the future. We wrote this book with the aim to encourage students and scientists to bring their efforts into this field. The book has been intended as written for the generalist at a level that could be understood by an undergraduate student in Physics, Chemistry or Biology or by scientists who are not in this field. We try to provide "first principles" introduction to each of the proposed approaches, exploring its attributes and hurdles along with clear working examples and real applications in a laboratory of chemistry. A part from giving a comprehensive view of the field, we try to indicate possible future developments, especially by considering the electrochemical and optical mediated detection methods that can make this technology amenable for integration into future analytical devices.

The Editors are sincerely thankful to all the authors of the chapters, without their passion for their work, this book would not have been possible. The Editors wish to formulate sincere thanks to Prof. Bruno

Pignataro, Full Professor in Physical Chemistry at the University of Palermo for having transferred the sincere passion for nanosciences and Physical Chemistry.

Dr. Arrabito acknowledges Prof. Christof Niemeyer, Professor of Chemical Biology at Karlsruhe Institute of Technology (KIT) and Director of the Institute for Biological Interfaces (IBG-1), for having permitted Dr. Arrabito to get interest DNA nanostructures and its applications in Biology. Dr. Arrabito wishes to send special thanks to Dr. Christian Falconi (Assistant Professor, University of Rome Tor Vergata) for generous support and possibility to work in the field of electronic biosensors. Dr. Arrabito also wants to thank Prof. Francesco Ricci (Associate Professor, University of Rome Tor Vergata) for the encouragements and the ideas which motivated Dr. Arrabito to consider DNA aptamers as potential breakthrough in biosensors field.

Dr. Wang would like to acknowledge Prof. Chunhai Fan for providing her consistent direction, motivation and the opportunity to work in one of the most advanced groups that deals with DNA nanotechnology. Dr. Wang also wants to thank Dr. Piero Morales, who has introduced her into DNA nanotechnology and teaching her through all her PhD years.

Without efforts and encouragements of the above cited professors and the invaluable contribution of all the authors of the book, nothing of this would have not been possible. And maybe new generation of scientists would not have got any idea of how great DNA nanoscience can be!

Giuseppe Arrabito
Department of Physics and Chemistry,
University of Palermo, Viale delle Scienze,
Parco d'Orleans II, 90128 Palermo, Italy

Liqian Wang
Laboratory of Physical Biology,
Shanghai Institute of Applied Physics,
Chinese Academy of Sciences,
Shanghai 201800, China

About the Authors

Giuseppe Arrabito received his PhD (2012) from Scuola Superiore di Catania. He carried out Post-Doctoral fellowships in the group of C.M. Niemeyer (2013) and in the group of C. Falconi (2014). He is currently a Post-Doctoral Scientist in the group of B. Pignataro. His main research interests are in the fields of bioanalytical chemistry, micro- and nano-biological array fabrication, Dip-Pen Nanolithography, DNA-directed material assembly for nanotechnological applications and ZnO nanostructures for biosensing.

Liqian Wang received her PhD degree (2016) in Materials Science from the University of Rome Tor Vergata. She worked as a visiting researcher (2014) in the group of Ilia N. Ivanov at Oak Ridge National Laboratory. She is currently a Post-Doctoral Scientist in the group of Chunhai Fan. Her research interests include the application of DNA nanostructures as nanoelectronic devices and the interface between organic self-assembly and inorganic nanotechnology.

Contents

Chapter 1

Nanotechnology and the Unique Role of DNA

Elisa-C. Schöneweiß, Andreas Jaekel and Barbara Saccà*

Centre for Medical Biotechnology (ZMB)
University of Duisburg-Essen
45117 Essen, Germany
*barbara.sacca@uni-due.de

This chapter is a clear introduction to the role of DNA molecule in Nanotechnology. A part from describing the canonical and alternative structures, this chapter is aimed at giving fundamentals of self-assembling strategy of DNA nanostructures.

1.1 Introduction

"What I want to talk about is the problem of manipulating and controlling things on a small scale." With these words Richard Feynman introduced his talk at Caltech on December 29th 1959, entitled *"There is plenty of Room at the Bottom"* [1], which signed the beginning of a new field of science: nanotechnology. In his visionary speech, the Nobel Laureate in

Physics anticipated the tremendous possibilities offered by the *"staggeringly small world that is below,"* some of which became reality only recently. For the first time, scientists were challenged to overstep the minimal size of things that can be manipulated, bringing miniaturization to the smallest possible level. He proposed for example to use light to arrange molecules or even single atoms in a desired pattern on a suitable surface. This idea found later application in a series of top-down approaches, such as photolithography, inject-printing and other micropatterning methods [2–4], all sharing the same principle: the exploitation of the laws of physics to pack a huge amount of information on an exceedingly small space. On the other hand, the opposite process, that is, the extraction of all possible information from a tiny object, was also envisioned as a realistic goal, achievable for example with the development of more powerful microscopes. Nowadays, complex molecular assemblies can be observed almost at atomic resolution thanks to the improvement of electron microscopy [5–7] and kinetic processes can be resolved with picosecond accuracy using advanced single-molecule techniques [8–11].

Storing and using the information carried in a tiny space is the way nature performs its tasks. Fundamental biological processes such as cellular duplication, growth, movement and interactions are all tightly regulated in a microscopic and crowded area. What is mostly striking about it is that the information driving the whole machinery is contained in a few micrometer-large fraction of the cell, the nucleus, in the form of an almost two meter-long chain, the DNA molecule, which carries about one bit of information every 50 atoms. In this sense, the DNA molecule is probably the most brilliant example of nanotechnology in nature, being itself a nanosized object, capable to store and transmit the information necessary to life in a simple and efficient way. It is therefore not surprising that the beginning of the nanotechnology era started only few years after the discovery of the structure and function of DNA by Watson and Crick in 1953 [12, 13] and that latest advancements in the field have been made using DNA as the material.

The idea of employing DNA to construct molecular objects was pioneered by Ned Seeman in the early 1980s [14] and since then structural DNA nanotechnology has rapidly evolved, both in terms of design principles and applications [15]. Using a simple four bases code and applying predictable base pairing rules, this bottom-up approach allows the realization of DNA objects of almost any desired shape and pattern, thus enabling to reach the nanosized world from the molecular level, with sub-nanometer resolution. A notable breakthrough in the field occurred in 2006, when Paul Rothemund published the first scaffolded DNA-origami approach [16]. This method revolutionized our vision of DNA as a construction material, giving accessibility to nanosized objects with a level of sophistication previously only hardly imaginable [17, 18]. Recently other design strategies have emerged [19–22], enlarging the spectrum of attainable structures and demonstrating that the future of DNA design is still bright. This, together with the availability of user-friendly design tools and synthetic accessibility of oligonucleotides at relatively low costs, enabled the exponential growth of DNA nanotechnology in the past few years [23].

Besides solving design challenges, scientists rapidly succeeded in demonstrating the use of those structures for realistic applications, from the development of addressable molecular pegboards for protein patterning [24–27] or encapsulation [28–32], to optoelectronic hybrid materials [33] and organic catalysts [34]. Another field in great expansion is coupled to the advancement of single-molecule technologies, enabling for example the precise localization and counting of molecules in spatially distributed samples or the disclosure of anomalous kinetic events occurring on a time scale normally not accessible by standard methods [8, 35–40]. Recently, DNA nanostructures have also shown interesting properties in cells [41] and are currently explored as promising targeting and delivery systems [42, 43]. For all these applications, the nucleic acid structure must be previously functionalized, i.e. chemically modified with small molecules or proteins [44]. At this purpose, not only classical DNA-conjugation strategies have been implemented and revisited [45], but other methods based on sequence-specificity recognition [46] and protein adaptors [47, 48] have been also successfully applied.

In addition to the use of DNA as a static framework for controlling the spatial arrangement of molecules, the programmability of nucleic acids structures can be conveniently employed to develop dynamic systems, whose interconversion between distinct species occurs in a predictable manner [49–51]. This principle has been successfully employed to develop bioinspired DNA-nanomotors [52, 53] and bioinformatic tools for emulating complex reaction circuits [54].

Finally, one should not forget the enormous contribution to the field of DNA nanotechnology given by Chad Mirkin, who firstly employed the DNA molecule as a recognition motif attached over the surface of metal nanoparticles for biosensing applications [55, 56]. Later, the same principle has been applied for linking distinct nanoparticles together, leading to the generation of colloidal nanocrystals [57, 58] and nanoplasmonic materials with advanced optical properties [59].

In conclusion, although many of the initial Feynman's predictions became true, one cannot but remaining astonished by the enormous progresses made in the field and the rapid advancement of technologies and design strategies to approach the nanosized world with such a high level of control. In this sense, the DNA molecule represents an ideal candidate for the realization of those self-organizational systems initially envisaged in Feynman's lecture, which can help us to manipulate matter and understand the basic principles of self-assembly in natural and man-made architectures.

1.2 The DNA Molecule

1.2.1 *The double helical B-form*

The fundamental building block of every DNA nanostructure is the DNA double helix. Understanding the structure and chemical–physical properties of the double helix is therefore necessary when the purpose is to design ordered assemblies of single DNA units and using them for further functionalization.

DNA is a polymer composed by the periodic repetition of four possible nucleotides (Figure 1.1). Each nucleotide is made up of three components: a phosphate group, a five-carbon sugar deoxyribose and a nitrogenous base (also termed nucleobase or simply base). The four nucleobases are adenine (A), guanine (G), thymine (T) and cytosine (C). Whereas the sugar–phosphate moieties provide the backbone of the nucleic acid structure, the four bases are responsible for its helical shape.

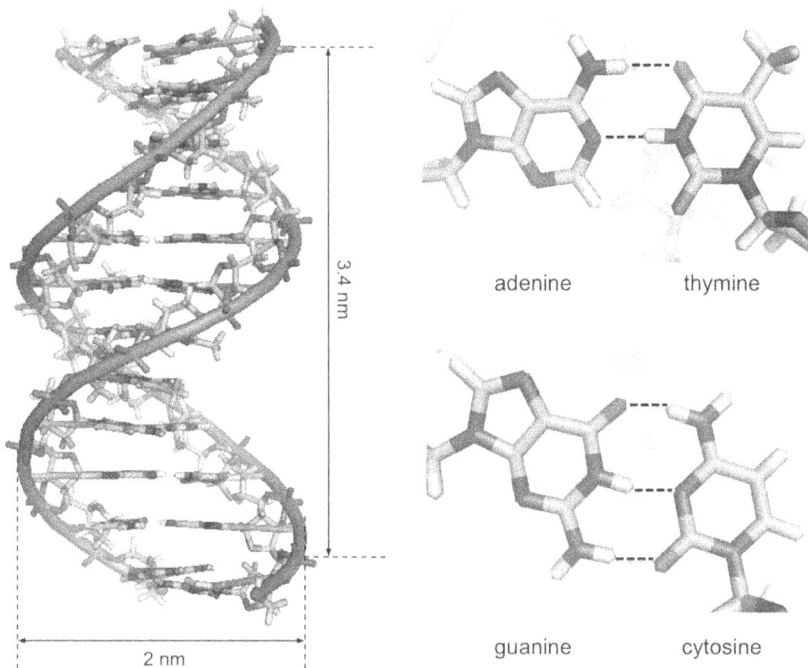

Figure 1.1. The DNA molecule in its most common B-form is a right-handed helical structure, which is 2 nm wide and rises up ca. 3.4 nm every helical turn (corresponding to ca. 10.5 base pairs (bp)). The four nucleobases (A, G, T, C) and their paring rule are also indicated. Note that whereas AT pairs are stabilized by two hydrogen bonds, GC pairs involve three bonds, which explains their higher contribution to helical stabilization. Figure modified from open sources (PDB: 1DUF).

This latter comes mostly from base stacking interactions among the aromatic nitrogenous bases. This is true for both single-stranded and double-stranded DNA. The double-stranded (ds) form, however, requires hydrogen bonding between the bases of each strand in order to form a stable double helix. The process is called *base pairing* because each nucleobase pairs only with a certain other nucleobase, named "complement": A pairs with T and G pairs with C. Such a Watson–Crick base-pairing rule is the fundamental law of each DNA design: once given a certain DNA sequence, its complementary sequence is uniquely defined. Such a reliable programmability of DNA nanostructures basically represents the main reason for the rapid and successful implementation of structural DNA nanotechnology.

In its B-form, the DNA is a right-handed double helix, about 2 nm wide and with a helical pitch of 3.4 nm comprising ca. 10.5 bp. The structural features of the glycosidic bonds of two pairing strands cause them to twist around the helical axis forming two unequally spaced grooves, a major groove (2.2 nm wide) and a minor groove (1.2 nm wide).

Although B-DNA is the most predominant form in cells, at least other two helical conformations are possible, A-DNA and Z-DNA, with strongly different geometry and dimensions. Whereas A-DNA mainly occurs in non-physiological conditions and adopts a wider (2.3 nm) and shorter helical pitch (ca. 2.8 nm/turn), the Z-form may appear upon DNA methylation and is a left-handed helix with a narrow diameter (1.8 nm) and a high helical pitch (4.6 nm/turn). Because of its abundance and stability, the B-DNA is the most widely form employed for construction of DNA nanostructures.

1.2.2 *Alternative DNA forms*

Non-canonical DNA motifs of particular importance in DNA nanotechnology are also hairpin loops, G-quadruplexes, i-motifs and triplex DNA.

Hairpin structures are formed by intra-strand base pairing, occurring when the sequence contains two inverted repeats, also termed palindromes (Figure 1.2(a)) [60]. The stability of this motif depends on the length of the double-helical segment (also termed stem), its sequence, as well as from the loop. This is the single-stranded region of the motif, which allows the sequence to fold back on itself to form the stem. Optimal loop lengths are about 4 to 8 bases long. Hybridization of the hairpin motif with a complementary sequence, which is also forming a hairpin, leads to a longer and therefore more stable double-helical segment. This conformational change can be easily programmed and in certain conditions even reversed, thus making hairpin motifs attractive structural features for dynamic DNA devices. Recently, the organized tethering of a distinct number of hairpin loops at precise locations of a DNA nanostructure has been advantageously used for actuation of large-scale translations [53].

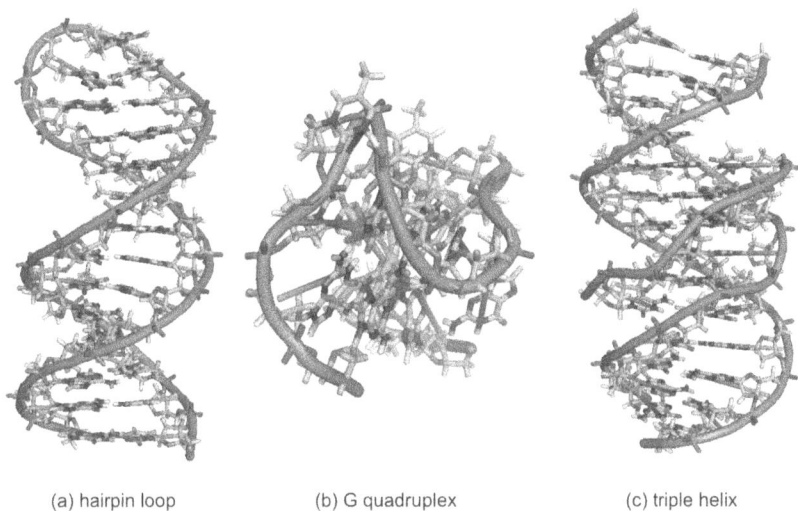

(a) hairpin loop   (b) G quadruplex   (c) triple helix

Figure 1.2. Alternative DNA motifs, largely employed in DNA nanotechnology. (a) The hairpin motif, formed by a duplex stem connected at one end by a single-stranded loop, (b) a G-quadruplex structure, stabilized by Hoogsteen hydrogen bonds between guanine bases and (c) a triple helix, formed by an oligopyrimidine single-stranded DNA binding to the major groove of an oligopurine sequence of the target duplex DNA. Figures adapted by open sources (PDB: 1JVE, 1C35 and 1BWG, respectively).

Widely employed in DNA nanotechnology are also *G-quadruplexes* (G4) and i-motifs. G4 structures are based on the stacking of several so-called G-quartets, which consist each of four guanine bases held together by Hoogsteen-type hydrogen bonding (Figure 1.2(b)) [61]. G4 structures are further stabilized by the presence of cations (generally sodium and potassium) in the central channel of the helix. A G4 motif can be formed by one or several DNA strands, oriented either in a parallel or antiparallel fashion. The structural diversity, folding properties and stabilities of G-quadruplex DNA have been extensively studied, both as a model for a non-canonical secondary DNA structure and as a pharmacological target for small molecules that have potential to impact gene expression [61].

i-Motifs are four-stranded DNA secondary structures which can form in cytosine rich sequences [62]. Stabilized by acidic conditions, they are made of two DNA duplexes held together in an antiparallel orientation by intercalated, cytosine–cytosine$^+$ bps. By virtue of their pH dependent folding (in slightly acidic conditions), i-motif forming DNA sequences have been used extensively as pH switches for applications in nanotechnology [63].

Finally, another non-conventional DNA structure is the triple helix, constituted by a single DNA strand, normally termed *triplex-forming oligonucleotide* (TFO), wrapped around a B-form double helix through Hoogsteen hydrogen bonds (Figure 1.2(c)). A TFO binds to the major groove of the target duplex DNA in a sequence-specific manner, generating base triplets with the purine bases. Therefore, formation of a triple-helix is limited by the presence of oligopyrimidine • oligopurine sequences in the DNA target. TFOs have been recently the focus of considerable interest in nanotechnology [64] for developing new molecular biology tools as well as diagnostic agents [65].

1.2.3 *Physical and chemical properties of the DNA*

Besides its predictable self-recognition properties, at least other two reasons have made the DNA molecule a supreme candidate as a construction material: (i) its flexibility (the double helix has a persistence length of 50 nm) and (ii) its easy synthetic accessibility and modification. The former enables to intertwine distinct DNA chains together into many different forms, thus enormously enlarging the spectrum of possible achievable shapes. The latter allows the production of nanostructures in short time and at low costs, offering the possibility to add desired chemical groups at selected positions of the DNA chain, for further functionalization at the single-molecule level.

1.3 DNA as Building Block of Self-assembled Nanostructures

The initial step in formation of self-assembled DNA nanostructures is the connection of distinct double helices together, creating so-called branched DNA motifs or DNA tiles. The simplest branched motif is the Holliday junction, a key intermediate in many types of genetic recombination and

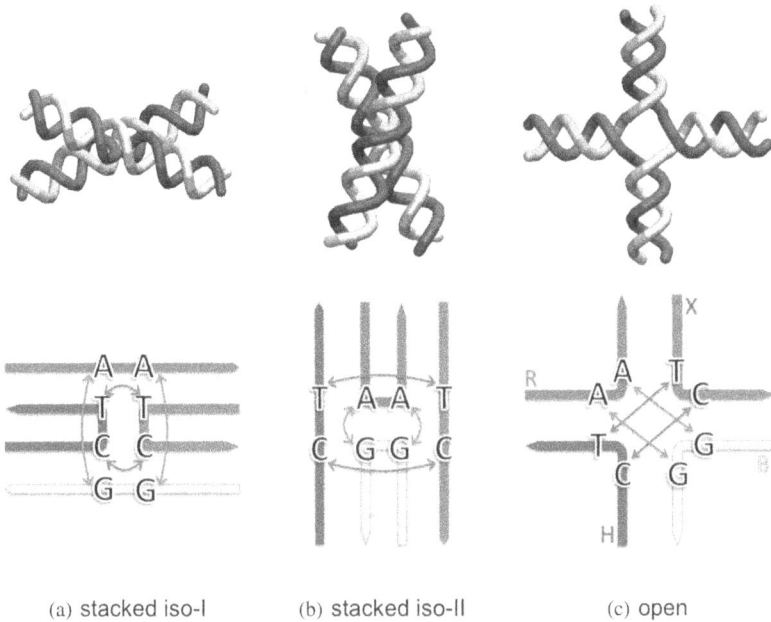

(a) stacked iso-I   (b) stacked iso-II   (c) open

Figure 1.3. The Holliday junction in its distinct conformations. The stacked iso-I (a) and iso-II (b) conformations predominate at modest and high salt concentrations. The open conformation (c) is mainly observed at low salt concentration, and also potentially acts as an intermediary of conformational changes between iso-I and II. Adapted with permission from Ref. [67].

ds break repair mechanisms. The Holliday junction forms when two homologous ds DNA molecules exchange their strands in one common branch point [66, 67]. Thus, the main structural feature of a Holliday junction is the symmetry of its component sequences, which causes the branch point to migrate (Figure 1.3).

Although this is of course essential for successful accomplishment of its biological function, it is instead deleterious for construction purposes, where a certain degree of structural rigidity of the constituent building blocks is desired. For this reason, initial efforts were focused on the development of an *immobile* Holliday junction, i.e. a motif which is designed such to avoid any topological isomerization and is therefore incapable to resolve into two distinct double-helical segments. This ambitious goal was achieved in 1982 by Ned Seeman and signed the beginning of structural

DNA nanotechnology [14]. In the following years, more complex DNA motifs have been developed and different design strategies have evolved, sometimes even against initial predictions, demonstrating that the field of DNA design may still reserve many surprises. A more detailed description of the current strategies and available software tools for design of DNA nanostructures will be given in Chapter 2.

Once the building blocks have been designed, the next step is to drive their self-assembly into the target structure. Different approaches are nowadays available to reach this purpose and are mainly related to the kind of "building units" chosen and the forces involved in their binding. Nevertheless, the underlying concept is equal in all cases: the entropic loss that occurs when a single element moves from an "unbound" state in solution to a "bound" state in the growing structure must be compensated by a gain in the energy of binding, thus favoring the self-assembly process. Intuitively, inter-unit recognition forces should be strong enough to ensure specificity, and therefore guide the system to a state of minimal energy, but sufficiently weak to enable reversibility of binding, such that incorrect bonds can disrupt. In this way, error-check and self-correction mechanisms can take place [68, 69]. If self-recognition interactions are too strong, the thermodynamic forces driving the process to the optimal bound state may be counteracted by the slowness of escaping from misbound structures, thus leading to kinetic trapping (Figure 1.4(a)).

Another advantage in using DNA as a construction material is the large variety of forces, which can be exploited to link DNA pieces together. Indeed, besides canonical and non-canonical hybridization forces (Figure 1.4(b)), *base stacking* offers another possibility to improve the binding force, particularly at the blunt ends of larger DNA shapes (Figure 1.4(c)) [70]. In this way, hierarchical self-assembly, i.e. the association of DNA units into higher-order architectures may be designed, leading in some cases even to macroscopic structures. Interactions between DNA units can be also directionally driven taking advantage of geometric factors, like the *shape-complementarity* of the binding surfaces (Figure 1.4(d)) [71]. Finally, long-range assemblies can also be favored by confining the random Brownian motion of the constituents units to a limited space, helping them to meet and bind [72–74]. Both approaches,

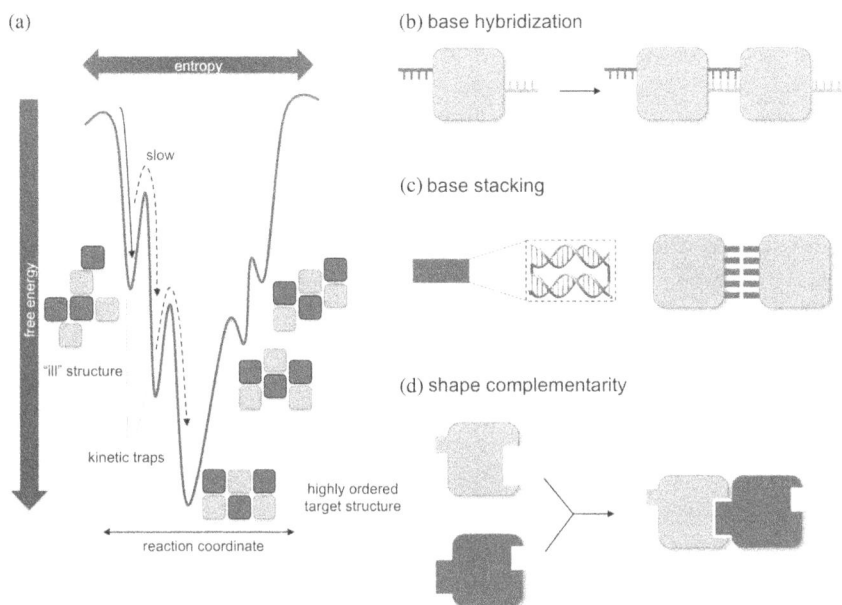

Figure 1.4. The main forces driving the self-assembly of DNA nanostructures. (a) A representative energy landscape of self-assembling DNA structures, showing that the step-wise hierarchical association of single building units may follow different paths. From the one side, short-life intermediate species may form which are capable to self-correct into the target highly-ordered structure. In other cases, the unit-to-unit interactions may be too strong to be counteracted, terminating the process into a kinetic trap, from which the ill structure cannot easily escape. Unit-to-unit interactions in DNA self-assembly are mainly hydrogen bonds between complementary bases (b). Other forces include base stacking at blunt end helices (c) or shape complementarity (d). Adapted with permission from Ref. [75].

i.e. shape-complementarity and *template-assisted growth*, may be viewed as a strategy to restrict the number of isoenergetic states, eventually leading to a narrower energy profile with a uniquely defined minimum (Figure 1.4(a)) [75].

1.4 Functionalization of DNA Nanostructures

Until now we have regarded the DNA molecule as the building unit of self-assembled nanostructures. However, to make such structures useful for realistic applications, chemical functionalities must be added. This can be done in two ways: from the one side, exploiting well-defined oligonu-cleotide sequences as substrates of highly specific DNA-modifying enzymes. From the other side, bringing desired DNA-tagged molecules at selected positions of the nanostructured template through hybridization with complementary single-stranded protruding arms. In both cases, the DNA acts as a recognition motif either towards proteins or complemen-tary DNA-tagged molecules. Specific examples of each strategy will be reported below.

1.4.1 *DNA as a recognition motif*

In the past few years, Sugiyama and coworkers developed a DNA nanochip for direct analysis of DNA-binding events. At this purpose, they immobi-lized DNA recognition elements in the central cavity of a frame-shaped nanostructure and analyzed binding or processing events at the single molecule level, mainly using high-speed atomic force microscopy. In this way they could control for example the tension of the immobilized dsDNA substrate and thus demonstrate the importance of DNA-strand relaxation and bending in base-excision repair mechanisms [76]. Similarly, by tethering a Holliday junction intermediate into the DNA frame in dif-ferent connection patterns, they found that the topological state of the initial DNA substrate dictates the outcome of the recombination process (Figure 1.5(a)) [77, 78].

In another important class of nanostructures, the DNA motif is not metabolized or processed, but simply recognized by the incoming pro-teins, thus serving as a molecular tag, for example, for targeting of tran-scription factors [29] or site-specific positioning of DNA-binding protein adaptors [79, 80]. A large family of functional DNA nanostructures relies on the use of DNA aptamers for biosensing applications. One of the most relevant studies of this kind was done by Yan and coworkers [81], who used a DNA platform to anchor two thrombin-binding aptamers targeting opposite sides of the same molecule (Figure 1.5(b)). By varying the dis-tance between the two ligands, they could demonstrate optimal bivalent binding at about 5.3 nm. This study represents the first example of using the spatial addressability of self-assembled DNA nanoscaffolds to control

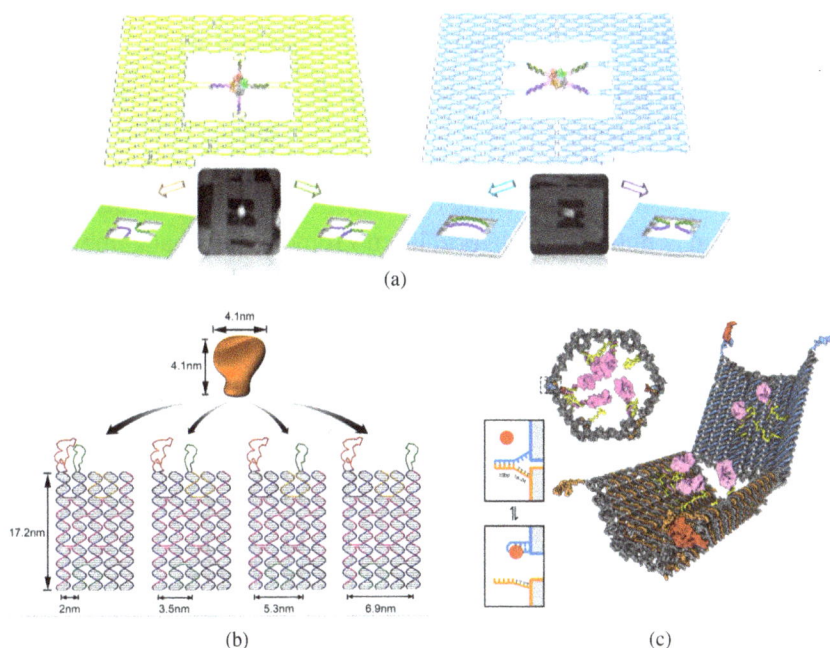

Figure 1.5. Examples of DNA nanostructures carrying additional DNA sequences as specific recognition motifs for DNA processing enzymes (a) and multivalent protein binding (b). DNA aptamers have been also used to actuate dynamic transformations (c), from a locked status (aptamer–complement binding) to an unlocked status (aptamer–protein binding). Adapted with permission from Refs. [32, 77, 81].

multi-component bimolecular interactions and to visualize them at a single-molecule level.

In other applications, DNA aptamers have been used to trigger dynamic transformations (Figure 1.5(c)). The principle behind the mechanical transformation is rather simple: two parts of the same construct are initially connected by an aptamer and its complementary strand. Addition of an aptamer-specific target induces the folding of the aptamer and displaces it from the duplex, thus unlocking the two parts previously bound together. This strategy has been used for the generation of mechanochemical sensing platforms [82], aptamer-encoded logic gates [32] or for controlled release of encapsulated cargo from a DNA cage [30].

1.4.2 Synthesis of DNA–protein conjugates

As mentioned above, besides employing DNA as a recognition motif for sequence-specific binding proteins, functionalization of DNA nanostructures can be attained through hybridization of DNA-tagged components to complementary strands protruding out of the nanoscaffold. Semisynthetic DNA–protein conjugates are hybrid biomacromolecules that combine the unique structure-directing properties of DNA with the almost unlimited variety of functional proteins. Proteins have been tailored by billions of years of evolution to perform highly specific functions, such as molecular recognition, catalytic turnover, energy conversion or translocation of other components across membranes. In this sense, DNA–protein conjugates represent an ideal tool for adding a desired functionality to nanostructures. Various methods for the coupling of synthetic DNA oligomers with proteins have been developed, based on either covalent or non-covalent strategies. The chemistry of DNA bioconjugation is too large to be treated here in detail and extensive literature is available on the topic [44, 83]. In the following, we will therefore give a brief survey of the most important methods adopted for the synthesis of DNA–proteins conjugates and how they have been used for functionalization of DNA nanostructures.

1.4.2.1 Covalent coupling

In the covalent coupling approach, a DNA strand is modified with a chemical function, which is then covalently bound to the protein-of-interest (POI). A typical approach is based on thiol chemistry between a thiol-modified DNA and genetically engineered cysteine residues of recombinant proteins, thus allowing to prepare DNA conjugates with defined stoichiometry and regioselectivity of the coupling site [84]. However, while disulfide bonds are still amenable to cleavage under reductive conditions, the use of crosslinkers bearing a maleimide functionality, such as sulfosuccinimidyl 4-(N-maleimidomethyl) cyclohexane-1-carboxylate (sulfoSMCC), enables irreversible coupling. SulfoSMCC is a representative example of the broad class of heterobispecific crosslinkers, which are used to couple amino-reactive groups exposed on the protein surface with thiol-modified DNA oligonucleotides [85]. If a

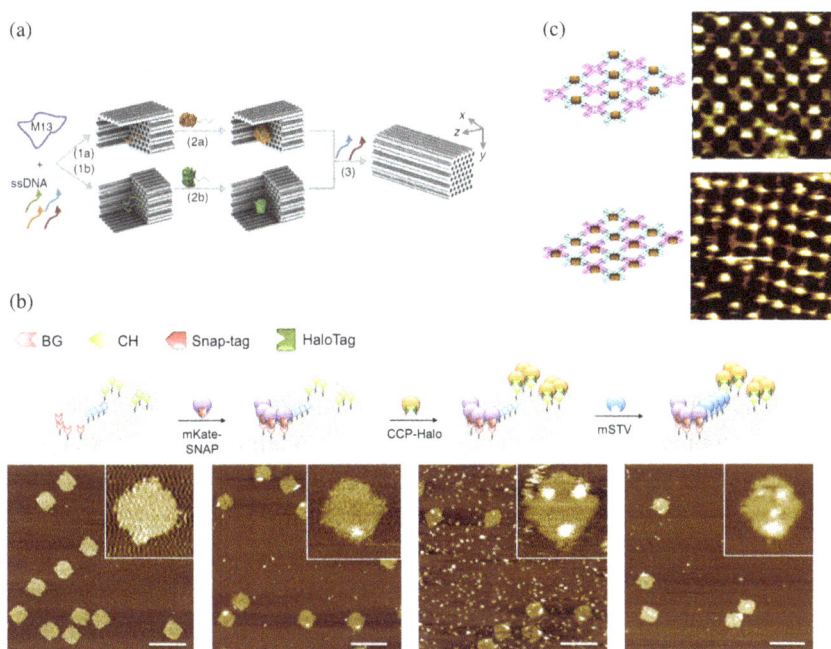

Figure 1.6. Examples of covalent (a) and (b) and non-covalent (c) DNA–protein complexes used to functionalize DNA nanostructures. Typically, DNA–protein conjugates are obtained by linking the two components by heterobifunctional crosslinkers (a). The conjugates are then attached to the nanostructure through hybridization with complementary strands. Other methods instead rely on the covalent binding of protein tags to suicide ligands previously tethered onto the nanostructure (b). Finally, the biotin–STV interaction is the most widely used non-covalent binding for decoration of DNA nanostructure surfaces (c). Scale bars are 100 nm. Adapted with permission from Refs. [26, 28, 89].

site-specific linkage is required and the POI contains accessible cysteine residues, the coupling order can be reversed, using amino-modified oligonucleotides [86]. The use of heterocrosslinkers is still one of the most largely employed methods for attachment of selected proteins to DNA nanostructures (Figure 1.6(a)) [28].

However, in many cases, it is not possible to engineer a single chemically accessible cysteine into the POI. Therefore, coupling methods are needed, which are orthogonal to the variety of functional groups in native proteins. A typical example is given by the self-labeling proteins

"Snap-tag" [87] and "HaloTag" [88] that specifically bind to their suicide ligands benzylguanin (BG) and chlorohexane (CH), respectively. Both Snap- and HaloTag have been genetically fused to POIs and subsequently reacted with BG- or CH-modified DNA oligonucleotides, allowing for the site-selective decoration of DNA nanostructures with multiple different proteins (Figure 1.6(b)) [89, 90].

1.4.2.2 *Non-covalent coupling*

Probably the most widely employed non-covalent interaction between a DNA strand and a protein takes advantage of the remarkable biomolecular recognition of D-biotin (vitamin H) by the homotetrameric proteins avidin or streptavidin (STV). Due to the enormous high affinity constant of the biotin-(strept)avidin interaction ($K_d = 10^{-15}$ M), the extreme chemical and thermal stability of the STV, and the commercial availability of biotinylated oligonucleotides, biotin–STV coupling has frequently been applied in structural DNA nanotechnology, either as a proof-of-principle for protein binding or as topographical marker for atomic force microscopy imaging (Figure 1.6(c)) [91]. Other affinity tags, such as the chromophore fluorescein (Fsc), can also be coupled to DNA oligonucleotides. Fsc can be used as a hapten to specifically bind immunoglobulin G (IgG) antibodies raised against it. For example, Mao and colleagues have used Fsc-derivatized oligonucleotides to assemble a hapten-modified DNA nanoarray, which was then used as template to assemble antibodies into high density arrays with a defined pitch of about 20 nm [26].

1.5 Dynamic DNA Motifs

The ultimate goal of DNA nanotechnology is the finest possible level of control over the *spatial* and *temporal* structure of matter. Thus, besides the use of DNA as a static framework to control the spatial arrangement of molecules, the programmability of nucleic acids structures can be conveniently employed to develop dynamic systems, whose interconversion between distinct species occurs in a predictable manner [49]. The first switchable molecular machine has been pioneered again by Ned Seeman and coauthors in 1999 [92]. In their work, they describe the

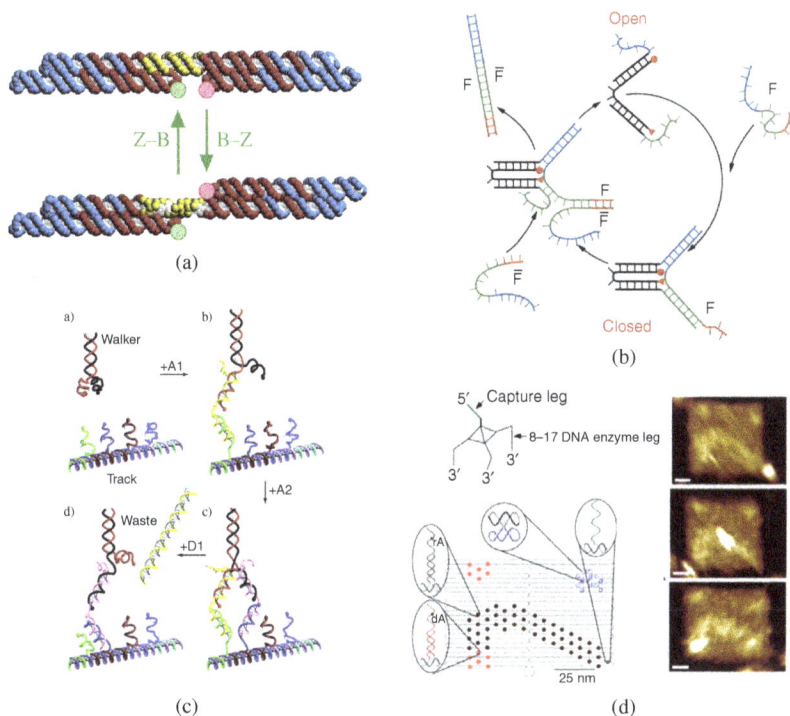

Figure 1.7. Examples of dynamic DNA devices, using the metal-induced B–Z transition of the DNA helix (a) or the well-established single-strand displacement mechanism (b). This latter has been used to create DNA-walkers, moving on 1D (c) or 2D (d) tracks along prescriptive directions. Adapted with permission from Refs. [93, 94, 97, 100].

realization of a supramolecular device consisting of two rigid double-crossover (DX) molecules connected by a short double-helical linker (Figure 1.7(a)). The screw motion of the device was triggered by the addition or removal of cobalt ions, which induced the transition from the B- to the Z-form of DNA. The operation has been monitored by fluorescence resonance energy transfer spectroscopy, measuring the relative proximity of two dye molecules attached to the free ends of the DX molecules.

One year later, Yurke and coworkers published the first DNA machine powered by DNA itself [93], thus establishing the basics of dynamic DNA nanotechnology. They reported the construction of a DNA device in which the DNA was used not only as a structural material, but also as a "fuel" (Figure 1.7(b)). The functioning of the machine was based on the predictable conformational transition between different species, driven by formation of thermodynamically more favored duplexes through a mechanism of strand invasion and single-strand displacement. In the following years, this principle has been successfully employed to develop bioinspired DNA-nanomotors [52] and bioinformatic tools for emulating complex reaction circuits [54]. Recently, Kuzuya and coauthors also demonstrated that the structural reconfiguration of large DNA origami can be precisely controlled and directed to one of three different states [94], thus suggesting that more complex systems can be designed, in which the interplay between structural and dynamic properties of the DNA molecule offers enormous potential for future applications.

The controlled nanomechanical actuation provided by strand displacement was also used to construct molecular devices that could continuously move along a predefined trajectory rather than switching between a limited number of fixed configurations (Figures 1.7(c) and 1.7(d)). In this way, so-called DNA walkers have been produced, which could move directionally along a DNA track, taking one step with every input added [95–99].

Many of these devices, however, require the addition of external input strands for continued operation and suffer from "poisoning" of the system due to formation of so-called duplex waste. As in macroscopic man-made machines, the efficiency of functioning is therefore limited. With the purpose to overcome this drawback, scientists developed alternative strategies, such as the single-strand displacement cascades. In these systems, the output of a reaction serves as the input to a downstream reaction, thus allowing for the engineering of complex autonomous devices [100, 101].

Finally, recent works have demonstrated the importance of kinetic control of strand displacement for the programmability of autonomous molecular machinery and molecular computation [102]. This can be achieved by insertion of mismatched bps, thus allowing to tune the energy landscape of the reaction in a predictable way. A remarkable example of the extreme degree of control achievable using DNA nanotechnology tools is represented by the work of Winfree and coauthors [103]. There, they showed how two well-defined frameworks, i.e. DNA tile self-assembly and DNA strand-displacement circuits, can be integrated to provide programmable kinetic control of self-assembly.

Thus, the future of DNA nanotechnology is without any doubt bright: integrating more complex structures with dynamic circuits one can envision to achieve a precise spatial and temporal organization of matter with a resolution of only few nanometers.

References

1. Feynman, R.P. (1960). There is plenty of room at the bottom. *Eng. Sci.* **23**, 22–36.
2. Zhao, J., Swartz, L.A., Lin, W.F., Schlenoff, P.S., Frommer, J., Schlenoff, J.B., and Liu, G.Y. (2016). Three-dimensional nanoprinting via scanning probe lithography-delivered layer-by-layer deposition. *ACS Nano.* **10**(6), 5656–5662.
3. Yuan, L.L., and Herman, P.R. (2016). Laser scanning holographic lithography for flexible 3D fabrication of multi-scale integrated nano-structures and optical biosensors. *Sci. Rep.* **6**, 22294.
4. Pal, R.K., Farghaly, A.A., Collinson, M.M., Kundu, S.C., and Yadavalli, V.K. (2016). Photolithographic micropatterning of conducting polymers on flexible silk matrices. *Adv. Mater.* **28**(7), 1406–1412.
5. Diebolder, C.A., Koster, A.J., and Koning, R.I. (2012). Pushing the resolution limits in cryo electron tomography of biological structures. *J. Microsc.* **248**(1), 1–5.
6. Schroder, R.R. (2015). Advances in electron microscopy: A qualitative view of instrumentation development for macromolecular imaging and tomography. *Arc. Biochem. Biophys.* **581**, 25–38.
7. Dubrovsky, A., Sorrentino, S., Harapin, J., Sapra, K.T., and Medalia, O. (2015). Developments in cryo-electron tomography for *in situ* structural analysis. *Arc. Biochem. & Biophys.* **581**, 78–85.
8. Tsukanov, R., Tomov, T.E., Liber, M., Berger, Y., and Nir, E. (2014). Developing DNA nanotechnology using single-molecule fluorescence. *Acc. Chem. Res.* **47**(6), 1789–1798.
9. Hildebrandt, L.L., Preus, S., and Birkedal, V. (2015). Quantitative single molecule FRET efficiencies using TIRF microscopy. *Faraday Discuss,* **184**, 131–142.

10. Schonfelder, J., De Sancho, D., and Perez-Jimenez, R. (2016). The power of force: Insights into the protein folding process using single-molecule force spectroscopy. *J. Mol. Biol.* **428**(21), 4245–4257.

11. Sustarsic, M., and Kapanidis, A.N. (2015). Taking the ruler to the jungle: Single-molecule FRET for understanding biomolecular structure and dynamics in live cells. *Curr. Opin. Struct. Biol.* **34**, 52–59.

12. Watson, J.D., and Crick, F.H. (1953). Genetical implications of the structure of deoxyribonucleic acid. *Nature.* **171**(4361), 964–967.

13. Watson, J.D., and Crick, F.H. (1953). Molecular structure of nucleic acids; a structure for deoxyribose nucleic acid. *Nature.* **171**(4356), 737–738.

14. Seeman, N.C. (1982). Nucleic acid junctions and lattices. *J. Theor. Biol.* **99**(2), 237–247.

15. Seeman, N.C. (2003). DNA in a material world. *Nature.* **421**(6921), 427–431.

16. Rothemund, P.W. (2006). Folding DNA to create nanoscale shapes and patterns. *Nature.* **440**(7082), 297–302.

17. Dietz, H., Douglas, S.M., and Shih, W.M. (2009). Folding DNA into twisted and curved nanoscale shapes. *Science.* **325**(5941), 725–730.

18. Douglas, S.M., Dietz, H., Liedl, T., Hogberg, B., Graf, F., and Shih, W.M. (2009). Self-assembly of DNA into nanoscale three-dimensional shapes. *Nature.* **459**(7245), 414–418.

19. Ke, Y., Ong, L.L., Shih, W.M., and Yin, P. (2012). Three-dimensional structures self-assembled from DNA bricks. *Science.* **338**(6111), 1177–1183.

20. Wei, B., Dai, M., and Yin, P. (2012). Complex shapes self-assembled from single-stranded DNA tiles. *Nature.* **485**(7400), 623–626.

21. Benson, E., Mohammed, A., Gardell, J., Masich, S., Czeizler, E., Orponen, P., and Hogberg, B. (2015). DNA rendering of polyhedral meshes at the nanoscale. *Nature.* **523**(7561), 441–444.

22. Zhang, F., Jiang, S., Wu, S., Li, Y., Mao, C., Liu, Y., and Yan, H. (2015). Complex wireframe DNA origami nanostructures with multi-arm junction vertices. *Nat. Nanotechnol.* **10**(9), 779–784.

23. Sacca, B., and Niemeyer, C.M. (2012). DNA origami: The art of folding DNA. *Angew. Chem. Int. Ed. Engl.* **51**(1), 58–66.

24. Yan, H., Park, S.H., Finkelstein, G., Reif, J.H., and LaBean, T.H. (2003). DNA-templated self-assembly of protein arrays and highly conductive nanowires. *Science.* **301**(5641), 1882–1884.

25. Chhabra, R., Sharma, J., Ke, Y., Liu, Y., Rinker, S., Lindsay, S., and Yan, H. (2007). Spatially addressable multiprotein nanoarrays templated by

aptamer-tagged DNA nanoarchitectures. *J. Am. Chem. Soc.* **129**, 10304–10305.

26. He, Y., Tian, Y., Ribbe, A.E., and Mao, C. (2006). Antibody nanoarrays with a pitch of approximately 20 nanometers. *J. Am. Chem. Soc.* **128**(39), 12664–12665.

27. Lin, C., Katilius, E., Liu, Y., Zhang, J., and Yan, H. (2006). Self-assembled signaling aptamer DNA arrays for protein detection. *Angew. Chem. Int. Ed. Engl.* **45**(32), 5296–5301.

28. Zhao, Z., Fu, J., Dhakal, S., Johnson-Buck, A., Liu, M., Zhang, T., Woodbury, N.W., Liu, Y., Walter, N.G., and Yan, H. (2016). Nanocaged enzymes with enhanced catalytic activity and increased stability against protease digestion. *Nat. Commun.* **7**, 10619.

29. Crawford, R., Erben, C.M., Periz, J., Hall, L.M., Brown, T., Turberfield, A.J., and Kapanidis, A.N. (2013). Non-covalent single transcription factor encapsulation inside a DNA cage. *Angew. Chem. Int. Ed. Engl.* **52**(8), 2284–2288.

30. Banerjee, A., Bhatia, D., Saminathan, A., Chakraborty, S., Kar, S., and Krishnan, Y. (2013). Controlled release of encapsulated cargo from a DNA icosahedron using a chemical trigger. *Angew. Chem. Int. Ed. Engl.* **52**(27), 6854–6857.

31. Zhang, C., Tian, C., Guo, F., Liu, Z., Jiang, W., and Mao, C. (2012). DNA-directed three-dimensional protein organization. *Angew. Chem. Int. Ed. Engl.* **51**(14), 3382–3385.

32. Douglas, S.M., Bachelet, I., and Church, G.M. (2012). A logic-gated nanorobot for targeted transport of molecular payloads. *Science.* **335**(6070), 831–834.

33. Acuna, G.P., Moller, F.M., Holzmeister, P., Beater, S., Lalkens, B., and Tinnefeld, P. (2012). Fluorescence enhancement at docking sites of DNA-directed self-assembled nanoantennas. *Science.* **338**(6106), 506–510.

34. Hung, Y.C., Bauer, D.M., Ahmed, I., and Fruk, L. (2014). DNA from natural sources in design of functional devices. *Methods.* **67**, 105–115.

35. Zhang, H., and Guo, P. (2014). Single molecule photobleaching (SMPB) technology for counting of RNA, DNA, protein and other molecules in nanoparticles and biological complexes by TIRF instrumentation. *Methods.* **67**, 169–176.

36. Johnson-Buck, A., and Walter, N.G. (2014). Discovering anomalous hybridization kinetics on DNA nanostructures using single-molecule fluorescence microscopy. *Methods.* **67**, 177–184.

37. Birkedal, V., Dong, M., Golas, M.M., Sander, B., Andersen, E.S., Gothelf, K.V., Besenbacher, F., and Kjems, J. (2010). Single molecule microscopy methods for the study of DNA origami structures. *Microsc. Res. Tech.* **74**(7), 688–698.

38. Helmig, S., Rotaru, A., Arian, D., Kovbasyuk, L., Arnbjerg, J., Ogilby, P.R., Kjems, J., Mokhir, A., Besenbacher, F., and Gothelf, K.V. (2010). Single molecule atomic force microscopy studies of photosensitized singlet oxygen behavior on a DNA origami template. *ACS Nano.* **4**, 7475–7480.

39. Rajendran, A., Endo, M., and Sugiyama, H. (2012). Single-molecule analysis using DNA origami. *Angew. Chem. Int. Ed. Engl.* **51**(4), 874–890.

40. Voigt, N.V., Torring, T., Rotaru, A., Jacobsen, M.F., Ravnsbaek, J.B., Subramani, R., Mamdouh, W., Kjems, J., Mokhir, A., Besenbacher, F., and Gothelf, K.V. (2010). Single-molecule chemical reactions on DNA origami. *Nat. Nanotechnol.* **5**(3), 200–203.

41. Bergamini, C., Angelini, P., Rhoden, K.J., Porcelli, A.M., Fato, R., and Zuccheri, G. (2014). A practical approach for the detection of DNA nanostructures in single live human cells by fluorescence microscopy. *Methods.* **67**, 185–192.

42. Okholm, A.H., Nielsen, J.S., Vinther, M., Sorensen, R.S., Schaffert, D., and Kjems, J. (2014). Quantification of cellular uptake of DNA nanostructures by qPCR. *Methods.* **67**, 193–197.

43. Ouyang, X., Li, J., Liu, H., Zhao, B., Yan, J., He, D., Fan, C., and Chao, J. (2014). Self-assembly of DNA-based drug delivery nanocarriers with rolling circle amplification. *Methods.* **67**, 198–204.

44. Sacca, B., and Niemeyer, C.M. (2011). Functionalization of DNA nanostructures with proteins. *Chem. Soc. Rev.* **40**, 5910–5921.

45. Dong, Y., Liu, D., and Yang, Z. (2014). A brief review of methods for terminal functionalization of DNA. *Methods.* **67**, 116–122.

46. Rusling, D.A., and Fox, K.R. (2014). Sequence-specific recognition of DNA nanostructures. *Methods.* **67**, 123–133.

47. Koussa, M.A., Sotomayor, M., and Wong, W.P. (2014). Protocol for sortase-mediated construction of DNA–protein hybrids and functional nanostructures. *Methods.* **67**, 134–141.

48. Ngo, T.A., Nakata, E., Saimura, M., Kodaki, T., and Morii, T. (2014). A protein adaptor to locate a functional protein dimer on molecular switchboard. *Methods.* **67**, 142–150.

49. Krishnan, Y., and Simmel, F.C. (2011). Nucleic acid based molecular devices. *Angew. Chem. Int. Ed. Engl.* **50**(14), 3124–3156.

50. Zhang, D.Y., and Seelig, G. (2011). Dynamic DNA nanotechnology using strand-displacement reactions. *Nat. Chem.* **3**(2), 103–113.

51. Feldkamp, U., and Niemeyer, C.M. (2008). Rational engineering of dynamic DNA systems. *Angew. Chem. Int. Ed. Engl.* **47**(21), 3871–3873.

52. Cheng, J., Sreelatha, S., Loh, I.Y., Liu, M., and Wang, Z. (2014). A bioinspired design principle for DNA nanomotors: Mechanics-mediated symmetry breaking and experimental demonstration. *Methods.* **67**, 227–233.

53. Sacca, B., Ishitsuka, Y., Meyer, R., Sprengel, A., Schoneweiss, E.C., Nienhaus, G.U., and Niemeyer, C.M. (2015). Reversible reconfiguration of DNA origami nanochambers monitored by single-molecule FRET. *Angew. Chem. Int. Ed. Engl.* **54**(12), 3592–3597.

54. Baccouche, A., Montagne, K., Padirac, A., Fujii, T., and Rondelez, Y. (2014). Dynamic DNA-toolbox reaction circuits: A walkthrough. *Methods.* **67**, 234–249.

55. Mirkin, C.A., Letsinger, R.L., Mucic, R.C., and Storhoff, J.J. (1996). A DNA-based method for rationally assembling nanoparticles into macroscopic materials. *Nature.* **382**(6592), 607–609.

56. Alivisatos, A.P., Johnsson, K.P., Peng, X., Wilson, T.E., Loweth, C.J., Bruchez, M.P., Jr., and Schultz, P.G. (1996). Organization of 'nanocrystal molecules' using DNA. *Nature.* **382**(6592), 609–611.

57. Nykypanchuk, D., Maye, M.M., van der Lelie, D., and Gang, O. (2008). DNA-guided crystallization of colloidal nanoparticles. *Nature.* **451**(7178), 549–552.

58. Park, S.Y., Lytton-Jean, A.K., Lee, B., Weigand, S., Schatz, G.C., and Mirkin, C.A. (2008). DNA-programmable nanoparticle crystallization. *Nature.* **451**(7178), 553–556.

59. Tan, S.J., Campolongo, M.J., Luo, D., and Cheng, W. (2011). Building plasmonic nanostructures with DNA. *Nat. Nanotechnol.* **6**(5), 268–276.

60. Chou, S.H., Chin, K.H., and Wang, A.H. (2003). Unusual DNA duplex and hairpin motifs. *Nucleic Acids Res.* **31**(10), 2461–2474.

61. Mendoza, O., Bourdoncle, A., Boule, J.B., Brosh, R.M., Jr., and Mergny, J.L. (2016). G-quadruplexes and helicases. *Nucleic Acids Res.* **44**(5), 1989–2006.

62. Day, H.A., Pavlou, P., and Waller, Z.A.E. (2014). i-Motif DNA: Structure, stability and targeting with ligands. *Bioorganic Med. Chem.* **22**(16), 4407–4418.

63. Dong, Y., Yang, Z., and Liu, D. (2014). DNA nanotechnology based on i-motif structures. *Acc. Chem. Res.* **47**(6), 1853–1860.

64. Yamagata, Y., Emura, T., Hidaka, K., Sugiyama, H., and Endo, M. (2016). Triple helix formation in a topologically controlled DNA nanosystem. *Chemistry.* **22**(16), 5494–5498.

65. Duca, M., Vekhoff, P., Oussedik, K., Halby, L., and Arimondo, P.B. (2008). The triple helix: 50 years later, the outcome. *Nucleic Acids Res.* **36**(16), 5123–5138.

66. Lilley, D.M. (2010). The interaction of four-way DNA junctions with resolving enzymes. *Biochem. Soc. Trans.* **38**(2), 399–403.
67. Wang, W., Nocka, L.M., Wiemann, B.Z., Hinckley, D.M., Mukerji, I., and Starr, F.W. (2016). Holliday junction thermodynamics and structure: Coarse-grained simulations and experiments. *Sci. Rep.* **6**, 22863.
68. Whitesides, G.M., and Grzybowski, B. (2002). Self-assembly at all scales. *Science.* **295**(5564), 2418–2421.
69. Whitesides, G.M., Mathias, J.P., and Seto, C.T. (1991). Molecular self-assembly and nanochemistry: A chemical strategy for the synthesis of nanostructures. *Science.* **254**(5036), 1312–1319.
70. Woo, S., and Rothemund, P.W. (2011). Programmable molecular recognition based on the geometry of DNA nanostructures. *Nat. Chem.* **3**(8), 620–627.
71. Gerling, T., Wagenbauer, K.F., Neuner, A.M., and Dietz, H. (2015). Dynamic DNA devices and assemblies formed by shape-complementary, non-base pairing 3D components. *Science.* **347**(6229), 1446–1452.
72. Woo, S., and Rothemund, P.W. (2014). Self-assembly of two-dimensional DNA origami lattices using cation-controlled surface diffusion. *Nat. Commun.* **5**, 4889.
73. Kocabey, S., Kempter, S., List, J., Xing, Y., Bae, W., Schiffels, D., Shih, W.M., Simmel, F.C., and Liedl, T. (2015). Membrane-assisted growth of DNA origami nanostructure arrays. *ACS Nano.* **9**(4), 3530–3539.
74. Aghebat Rafat, A., Pirzer, T., Scheible, M.B., Kostina, A., and Simmel, F.C. (2014). Surface-assisted large-scale ordering of DNA origami tiles. *Angew. Chem. Int. Ed. Engl.* **53**(29), 7665–7668.
75. Pfeifer, W., and Saccà, B. (2016). From nano to macro through hierarchical self-assembly: The DNA paradigm. *Chembiochem.* **17**, 1063–1080.
76. Endo, M., Katsuda, Y., Hidaka, K., and Sugiyama, H. (2010). A versatile DNA nanochip for direct analysis of DNA base-excision repair. *Angew. Chem. Int. Ed. Engl.* **49**(49), 9412–9416.
77. Suzuki, Y., Endo, M., Katsuda, Y., Ou, K., Hidaka, K., and Sugiyama, H. (2014). DNA origami based visualization system for studying site-specific recombination events. *J. Am. Chem. Soc.* **136**(1), 211–218.
78. Suzuki, Y., Endo, M., Canas, C., Ayora, S., Alonso, J.C., Sugiyama, H., and Takeyasu, K. (2014). Direct analysis of Holliday junction resolving enzyme in a DNA origami nanostructure. *Nucleic Acids Res.* **42**(11), 7421–7428.
79. Ngo, T.A., Nakata, E., Saimura, M., and Morii, T. (2016). Spatially organized enzymes drive cofactor-coupled cascade reactions. *J. Am. Chem. Soc.* **138**(9), 3012–3021.

80. Nakata, E., Liew, F.F., Uwatoko, C., Kiyonaka, S., Mori, Y., Katsuda, Y., Endo, M., Sugiyama, H., and Morii, T. (2012). Zinc-finger proteins for site-specific protein positioning on DNA-origami structures. *Angew. Chem. Int. Ed. Engl.* **51**(10), 2421–2424.

81. Rinker, S., Ke, Y., Liu, Y., Chhabra, R., and Yan, H. (2008). Self-assembled DNA nanostructures for distance-dependent multivalent ligand-protein binding. *Nat. Nanotechnol.* **3**(7), 418–422.

82. Koirala, D., Shrestha, P., Emura, T., Hidaka, K., Mandal, S., Endo, M., Sugiyama, H., and Mao, H. (2014). Single-molecule mechanochemical sensing using DNA origami nanostructures. *Angew. Chem. Int. Ed. Engl.* **53**(31), 8137–8141.

83. Hermanson, G.T. Chapter 1 — Introduction to Bioconjugation. In *Bioconjugate Techniques (Third edition)*, Academic Press: Boston, 2013; pp. 1–125.

84. Corey, D.R., and Schultz, P.G. (1987). Generation of a hybrid sequence-specific single-stranded deoxyribonuclease. *Science.* **238**, 1401–1403.

85. Niemeyer, C.M., Sano, T., Smith, C.L., and Cantor, C.R. (1994). Oligonucleotide-directed self-assembly of proteins: Semisynthetic DNA-streptavidin hybrid molecules as connectors for the generation of macro-scopic arrays and the construction of supramolecular bioconjugates. *Nucleic Acids Res.* **22**(25), 5530–5539.

86. Kukolka, F., and Niemeyer, C.M. (2004). Synthesis of fluorescent oligonu-cleotide-EYFP conjugate: Towards supramolecular construction of semi-synthetic biomolecular antennae. *Org. Biomol. Chem.* **2**, 2203–2206.

87. Keppler, A., Gendreizig, S., Gronemeyer, T., Pick, H., Vogel, H., and Johnsson, K. (2003). A general method for the covalent labeling of fusion proteins with small molecules *in vivo*. *Nat. Biotechnol.* **21**(1), 86–89.

88. Los, G.V., and Wood, K. (2007). The HaloTag: A novel technology for cell imaging and protein analysis. *Methods Mol. Biol.* **356**, 195–208.

89. Sacca, B., Meyer, R., Erkelenz, M., Kiko, K., Arndt, A., Schroeder, H., Rabe, K.S., and Niemeyer, C.M. (2010). Orthogonal protein decoration of DNA origami. *Angew. Chem. Int. Ed.* **49**, 9378–9383.

90. Meyer, R., and Niemeyer, C.M. (2011). Orthogonal protein decoration of DNA nanostructures. *Small.* **7**, 3211–3218.

91. Park, S.H., Yin, P., Liu, Y., Reif, J.H., LaBean, T.H., and Yan, H. (2005). Programmable DNA self-assemblies for nanoscale organization of ligands and proteins. *Nano Lett.* **5**(4), 729–733.

92. Mao, C., Sun, W., Shen, Z., and Seeman, N.C. (1999). A nanomechanical device based on the B-Z transition of DNA. *Nature.* **397**(6715), 144–146.

93. Yurke, B., Turberfield, A.J., Mills, A.P., Jr., Simmel, F.C., and Neumann, J.L. (2000). A DNA-fuelled molecular machine made of DNA. *Nature.* **406**(6796), 605–608.

94. Kuzuya, A., Watanabe, R., Hashizume, M., Kaino, M., Minamida, S., Kameda, K., and Ohya, Y. (2014). Precise structure control of three-state nanomechanical DNA origami devices. *Methods.* **67**, 250–255.

95. Gu, H., Chao, J., Xiao, S.J., and Seeman, N.C. (2010). A proximity-based programmable DNA nanoscale assembly line. *Nature.* **465**(7295), 202–205.

96. Lund, K., Manzo, A.J., Dabby, N., Michelotti, N., Johnson-Buck, A., Nangreave, J., Taylor, S., Pei, R., Stojanovic, M.N., Walter, N.G., Winfree, E., and Yan, H. (2010). Molecular robots guided by prescriptive landscapes. *Nature.* **465**(7295), 206–210.

97. Tomov, T.E., Tsukanov, R., Liber, M., Masoud, R., Plavner, N., and Nir, E. (2013). Rational design of DNA motors: Fuel optimization through single-molecule fluorescence. *J. Am. Chem. Soc.* **135**(32), 11935–11941.

98. Omabegho, T., Sha, R., and Seeman, N.C. (2009). A bipedal DNA Brownian motor with coordinated legs. *Science.* **324**(5923), 67–71.

99. Shin, J.S., and Pierce, N.A. (2004). A synthetic DNA walker for molecular transport. *J. Am. Chem. Soc.* **126**(35), 10834–10835.

100. Yin, P., Choi, H.M., Calvert, C.R., and Pierce, N.A. (2008). Programming biomolecular self-assembly pathways. *Nature.* **451**(7176), 318–322.

101. Green, S.J., Lubrich, D., and Turberfield, A.J. (2006). DNA hairpins: Fuel for autonomous DNA devices. *Biophys. J.* **91**(8), 2966–2975.

102. Machinek, R.R., Ouldridge, T.E., Haley, N.E., Bath, J., and Turberfield, A.J. (2014). Programmable energy landscapes for kinetic control of DNA strand displacement. *Nat. Commun.* **5**, 5324.

103. Zhang, D.Y., Hariadi, R.F., Choi, H.M., and Winfree, E. (2013). Integrating DNA strand-displacement circuitry with DNA tile self-assembly. *Nat. Commun.* **4**, 1965.

Chapter 2

DNA as Building Material at the Nanoscale: From Concepts to Software-aided Design

Wolfgang Pfeifer*, Georg Homa*, Giuseppe Arrabito[†]
and Barbara Saccà[*,‡]

*Centre for Medical Biotechnology (ZMB)
Universitätstr. 2 of Duisburg-Essen
45117 Essen, Germany
[†]Department of Physics and Chemistry
University of Palermo
Viale delle Scienze, Parco d'Orleans II, 90128 Palermo, Italy
[‡]barbara.sacca@uni-due.de

Chapter 2 is the most technical chapter of the book, since it deals with the different approaches and strategies allowing for design of DNA nano-structures such as tile-to-tile assembly or scaffold based design. The reader will also find information about DNA aptamers design for applications in analytical chemistry.

2.1 Basic Concepts of Design

Structural DNA nanotechnology is an example of how scientific original-
ity and human imagination can bring simple natural rules to such a high
level of sophistication to result into the founding of a new field of science.
Undoubtedly, the driving force of such a progress is DNA design. Since
the pioneering work of Ned Seeman in the early 1980s [1], multiple
design strategies have been developed. Despite all rely on the Watson–
Crick base-pairing rule, they explore different ways for folding DNA into
complex topologies.

The main reason for the rapid assessment of distinct design approaches
is the predictability of the DNA self-recognition interactions, which has a
remarkable consequence: once a sequence is given, its complementary
sequence is uniquely defined. In this sense, the design of a single double
helix is a rather simple exercise. The task however becomes more difficult
when the purpose is to generate a pool of DNA sequences, which self-
assemble into *distinct* linear duplexes. In this case, indeed, the availability
of only four bases gives rise to sequence subsets that are inevitably identical. To bypass this problem, these subsets should be as short and rare as
possible, such that the selected sequences will result maximally different.
This means that the probability of each strand to hybridize *exclusively*
with its complement will become maximal [2–4]. Fortunately, software
tools are nowadays available, which basically "translate" this sequence
diversity into a numerical quantity, thus allowing for ranking the sequence
candidates according to a desired property [5].

The next level of design difficulty appears when distinct DNA helices
must be designed, such to interlace one another at one or more points
(called junctions) forming intricate shapes of higher structural order,
termed tiles [6]. In the design of DNA tiles, not only geometric factors and
topological constraints become important but also the thermodynamic and
kinetic factors that regulate the interconversion between the transient species must be taken into account.

In the present chapter, we will describe how DNA tiles, and in general, DNA building units, can be designed from scratch and which parameters can be varied to tune their properties, such as the mechanical

flexibility, the global curvature and the strength of the interactions with other units to form extended structures. These latter represent the last level of design challenge, where a subtle balance between different forces must be achieved to ensure the successful building up of the desired assembly [7]. To achieve this purpose, the interactions between the binding units should be strong enough to be able to propagate over long distances but sufficiently weak to enable unbinding and self-corrections mechanisms, thus preventing kinetic trapping of misfolded structures [8, 9]. Besides base hybridization, other forces such as base stacking and shape complementarity [10–12], are often used to tune the interactions between the single DNA units of the system, allowing to direct the self-assembly process towards specific target shapes. In addition, experimental conditions such as temperature, annealing rate, ionic strength as well as the presence of surface supports or nucleation seeds can be optimized to improve the self-assembly yield by simply increasing the probability of the units to meet and bind [13–18].

It is therefore clear that the challenge of DNA design goes much beyond the simple base-complementarity rule and that robust theoretical models and software tools are necessary to enable the engineering and analysis of DNA structures. Fortunately, many of the instruments needed to achieve this goal are freely available in the internet and readily accessible to the scientific community, thus notably contributing to the assessment and rapid spread of this field of science.

In the following, we will give a brief introduction of the two main approaches used until now for designing DNA nanostructures and will describe the single strategies more in detail in the dedicated sections.

2.2 The Two "Schools" of DNA Design

Until now, the design of a DNA nanostructure can be performed following one of two possible schools of thought: (i) the so-called multi-stranded (or tile-based) approach, historically the first method established in the field of DNA nanotechnology (Figure 2.1(a)) [1] and (ii) the scaffold-based approach, commonly known as DNA origami, firstly reported by Rothemund in 2006 (Figure 2.1(b)) [19].

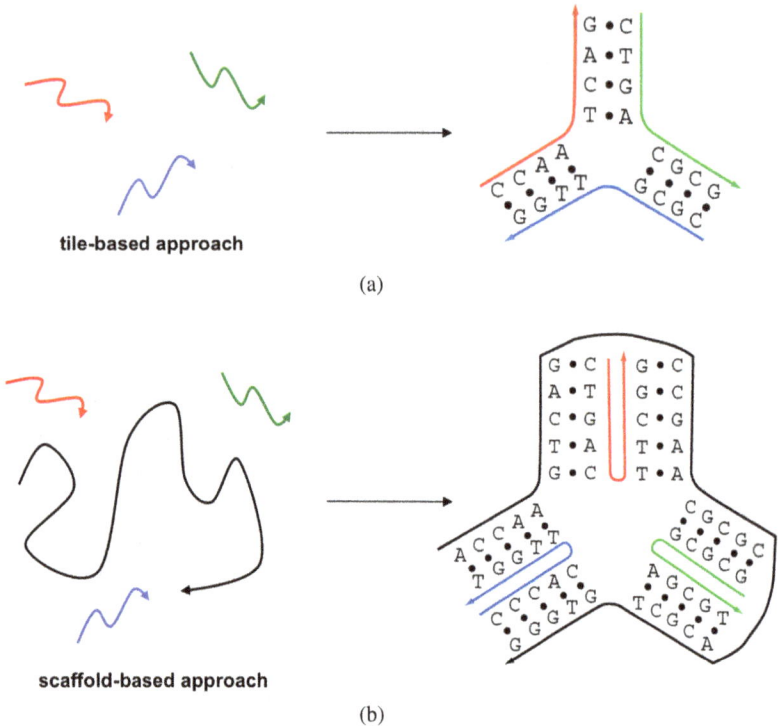

Figure 2.1. The two assembly strategies used in structural DNA nanotechnology. In the multi-stranded (or tile-based) approach (a), oligonucleotides are designed to self-assemble with each other into a well-defined branched DNA motif (also called "tile"). Hierarchical assembly of such motifs into large (finite or infinite) periodic matter is achieved through cohesion of sticky-ends. In the DNA-origami technique (b), a long single-stranded DNA is folded into a finite-sized shape by the help of many shorter oligonucleotide sequences, also called staple strands. Adapted with permission from Ref. [22].

The two strategies basically differ in the absence or presence of a long single-stranded sequence, named scaffold, as the "main actor" of the assembly process. This makes the design approaches, the attainable structures and experimental settings completely different and to some extent even divergent. In the multi-stranded procedure, DNA sequences are designed to self-assemble into branched motifs (also called tiles) of defined geometry [6], following a rather strict principle of sequence-symmetry minimization [5]. According to this principle, sequences participating in the formation of the motif are chosen to be maximally different from one another, such that competitive formation of alternative secondary structures will be minimized. In a further step, those motifs are used as building blocks for the construction of large periodic matter through cohesion of their sticky-ends. This requires a high control over the stoichiometry and purity of the constituent oligonucleotides, thus resulting in error-prone and lengthy synthetic processes. In the DNA-origami technique, instead, a long single-stranded DNA sequence (termed scaffold or template) is folded into a desired 2D or 3D shape by the help of up to hundreds of shorter oligonucleotides (also named staples). Although the first examples of such a scaffold-based approach were reported by Yan *et al.* [20] and Shih and coworkers [21], neither of these two papers had the impressive impact of Rothemund's publication [19].

The high performance of the DNA-origami method is mainly attributed to the entropic advantage in using a single long scaffold strand for folding [20]. In fact, since staple strands hybridize with a common scaffold rather than with each other, their relative stoichiometric ratio is no more relevant. Moreover, initial correct arrangement of the scaffold favors the correct binding of the remaining staples, such that possibly existing wrong or truncated sequences are easily displaced by strand invasion and exchange mechanisms. Consequently, experimental errors and time of synthesis are dramatically reduced because tight control over stoichiometry and purity of the oligonucleotides is no longer necessary. This, together with the capability to generate nanoobjects with complex shapes of pre-defined dimensions and full molecular addressability, makes the DNA origami a very robust and powerful tool for construction of DNA-based architectures [22].

Recently, alternative strategies have been also reported, belonging either to the multi-stranded or scaffold-based class. An example of tile-based approach is the wave tile design (Figure 2.2(a)), which uses two component strands interlaced back and forth to form flexible periodic architectures [23]. Using multi-arms junctions with low symmetry, intricate tessellation patterns based on Archimedean tiling have been also produced (Figure 2.2(b)) [24]. Aperiodic structures are instead more easily achievable through the so-called single-stranded tile (SST) approach (Figures 2.2(c) and 22.2(d)) that relies on the use of several hundreds of

(a)

(b)

(c)

(d)

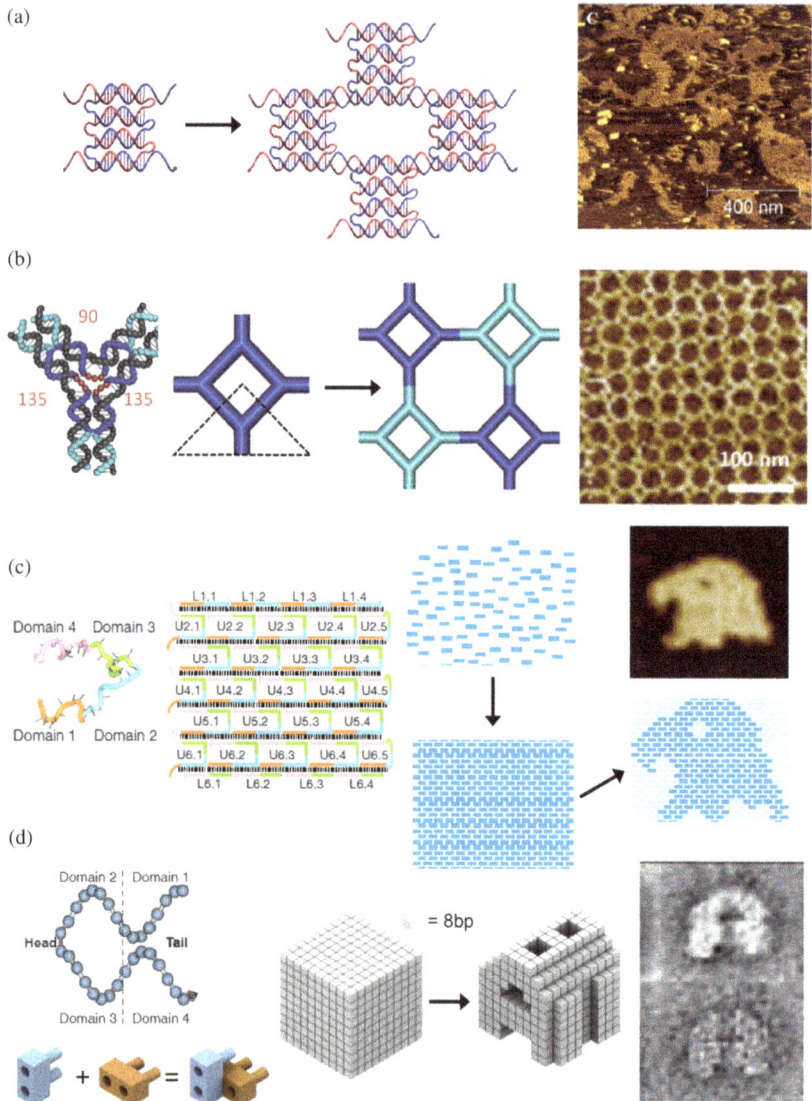

Figure 2.2. Examples of tile-based approaches. Whereas the weave tile design uses two component strands interlaced back and forth into flexible periodic architectures (a) [23], DNA branched junction of variable topological symmetry enable the formation of Platonic or Archimedean tessellation patterns (b) [24]. Instead, the SST approach relies on the use of several hundreds of distinct sequences designed to hybridize one another into simply concatenated duplexes, thus giving rise to 2D (c) or 3D (d) structures with a high level of complexity and addressability [25, 26]. Adapted with permission from Refs. [23–26].

distinct sequences designed to hybridize one another into concatenated duplexes [25, 26]. This method enables the generation of more than 100 arbitrary shapes from modular combinations of several hundreds distinct molecular pieces. Despite the low yield of the process and the challenging design of the sequences, the sophistication of the structures attained and their modular construction make this approach still unrivaled.

Using instead a scaffold routing algorithm based on graph theory [27] or wireframe patterns with controlled angles [28], net-like DNA shapes can be produced which would be otherwise almost impossible to achieve with other methods (see paragraph 1.4). Altogether, the diversity of current design procedures and the frequent emergency of novel approaches demonstrate the huge potential of DNA-guided self-assembly, where a simple rule applied to four pieces can still offer many possibilities.

2.3 The Multi-stranded Approach

2.3.1 *The first immobile junction designed by SEQUIN*

The first rationally designed DNA motif was an immobile Holliday junction [1], later optimized into the J1 motif (Figure 2.3) [5]. Contrarily to its analog transient species naturally occurring in genetic replication and recombination, the J1 junction was designed to avoid (or minimize) its topological isomerization, i.e. its resolution into distinct double-helical segments.

In this sense, the definition of immobile refers to the incapability of the motif to undergo branch migration at the junction and is not correlated to its structural properties. The J1 motif was designed with SEQUIN, a program to assign nucleic acid SEQUences INteractively. More correctly, the SEQUIN program was developed in order to design the prototype motif of structural DNA nanotechnology and is still one of the most dominant software tools for the design of immobile DNA junctions in general.

The first step in designing a DNA junction is to define its *molecular connectivity*. This implies specification of the number and chain length of the arms, presence and position of the junction, single-stranded loops and

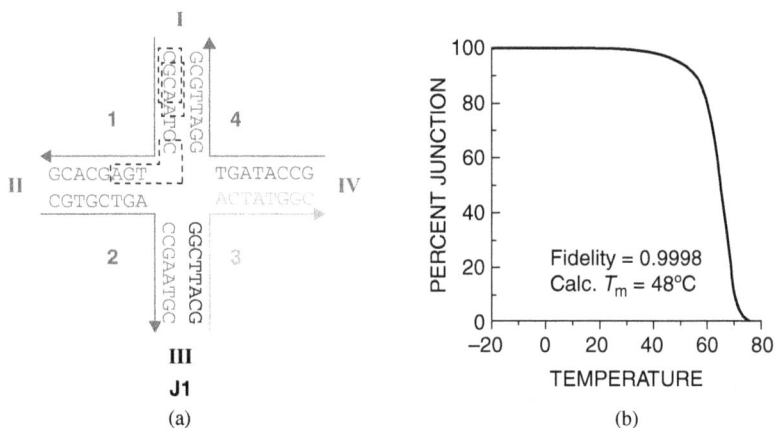

Figure 2.3. The optimized immobile DNA junction (J1). (a) The structure is composed by four hexadecamers forming a cross-like structure with 8 bp long arms. The structure has been designed with the SEQUIN program according to a sequence-symmetry minimization algorithm, which minimizes the probability of formation of alternative assemblies. Some critons are indicated by dashed boxes. (b) Besides sequence symmetry rules, energetic criteria (i.e. fidelity and melting profile) must be also taken into consideration for the realization of an immobile junction. The J1 motif has been selected as the "best" candidate of a four-arm junction, because presenting the highest calculated melting temperature among those motifs with the highest fidelity and a sigmoidal (all-or-none) transition profile. Figure (b) was adapted with permission from Ref. [5].

double-stranded linkages between the arms. In terms of SEQUIN, the *rank* (*R*) of a structure is defined as the number of double-helical portions, which directly emerge from the junction. Therefore, in the J1 motif of Figure 2.3(a), the four-arms junction has rank $R = 4$. Another structural parameter of the junction is the *bend*: i.e. the phosphodiester linkage, which is flanked by bases paired to different strands. In the example shown in Figure 2.3(a), the bend of strand 1 in the final structure is flanked by one C and one T, respectively paired to one G (of strand 4) and one A (of strand 2). Similarly, the other three strands also show one bend; the target four-arm junction has therefore totally four bends. Each of the four strands participating in the formation of the J1 motif is 16 bases long and can be considered as composed of 13 consecutive and partially overlapping four-base segments (highlighted by dashed boxes in strand 1 of Figure 2.3(a)). In the reported example, the first segment at the 5′ terminus of strand 1 will be the CGCA tetramer, the second segment (translated only by one nucleobase position towards the 3′ direction) will be therefore the GCAA sequence and so on, up to the last four-bases long segment at the 3′ end of the same strand, which will be CACG. Each of these segments is termed a *criton*. The number of different critons of length N generated by a four-bases code is 4^N. Thus, the total number of four-bases long critons ($N = 4$) will be $4^4 = 256$. These are the different subsequences that are in principle available for construction of the desired junction. At this point a selection criterion must be defined in order to reduce the pool of available sequences. In the SEQUIN program such criterion is the minimization of the sequence symmetry. Basically, the software tries to avoid base complementarity at undesired positions, thus reducing the chances of formation of alternative associations. More in detail, the program checks for the fulfillment of the following requisites within all selected pairing regions: (i) each criton may appear only once throughout the whole structure; (ii) the complement to any criton that spans a bend must not be present in any strand; (iii) self-complementary critons are not permitted; (iv) the same base pair at the junction can appear only twice and, if so, must be located on adjacent arms. In the J1 motif of Figure 2.3(a), the following base pairs are present in the central junction: CG (north arm), TA (east), GC (south) and AT (west). These are the four possible *distinct* combinations allowed, when considering the polarity of each double-helical arm. For example, in the CG pair of the north arm, the C occupies position 8 (counting from the 5′ terminus of strand 1) and G occupies position 9 of the complementary strand 4. This base-pairing configuration differs from the GC pair of the south arm, in which the two bases exchange their relative position (G is now in position 8 and C in position 9). This opposite polarity prevents the hybridization of the bases located in opposite arms.

If a rule violation (e.g. use of a self-complementary criton) is necessary because of additional requirements (e.g. the placement of a restriction site) the user can choose to tolerate it.

In a further step, free-energy criteria are taken into account to ensure reliable attainment of a stable branched motif at working temperature conditions. The program addresses this issue by adding two levels of design control: (i) calculation of the *fidelity* of the junction and (ii) calculation and plotting of the melting curves. The fidelity (*p*) of a structure is described as a probability function according to the Boltzmann's distribution:

$$p = \frac{e^{\frac{-\Delta G_J}{RT}}}{Z},$$

where ΔG_J is the free-energy of the desired junction, R is the gas constant and Z is the partition function, that is, the sum of the energetic contributions given by all competitive pairings represented by adjacent sets of two base pairs. The fidelity of a junction can be calculated from the $\Delta G°$ values reported in the literature for pairing of two-bases long segments [29], assuming that the equilibrium constant for junction formation ($K_J = \exp -\Delta G_J/RT$) is given by the weighted product of the binding constants of all two bp-subunits (i.e. $K_J = \beta K_1 K_2 K_2 \ldots K_{n-1}$; where β is the nucleation constant for initial strand hybridization and n is the length of the complementary sequences (all values are intended at 25°C) [5]. The junctions with the highest fidelity values are retained by the program and further analyzed for their thermal stability. Theoretical melting curves for the DNA junctions are then calculated assuming equal initial strand concentration for all strands participating in tile assembly. Melting curves are plotted as fraction of the junction formed in function of the temperature. A well-designed junction should present a sharp uniphasic melting profile with maximal melting temperature (Figure 2.3(b)).

Experience shows that 100% stringent observation of the rules described above is not always necessary and that a certain degree of flexibility through introduction of manual changes from the user is often more efficient. The accuracy of the SEQUIN algorithm for the design of DNA motifs of desired geometry has been largely demonstrated by the successful realization of DNA junctions of higher complexity, such as five, six, eight and even twelve-arm junctions (Figures 2.4(a)–2.4(d)) [30, 31]. The program has also been used for building motifs with multiple branch points, such as the double-crossover (DX) molecules (Figures 2.4(e) and 2.4(f)) [32] or the 4 × 4 tile (Figure 2.4(g)) [33].

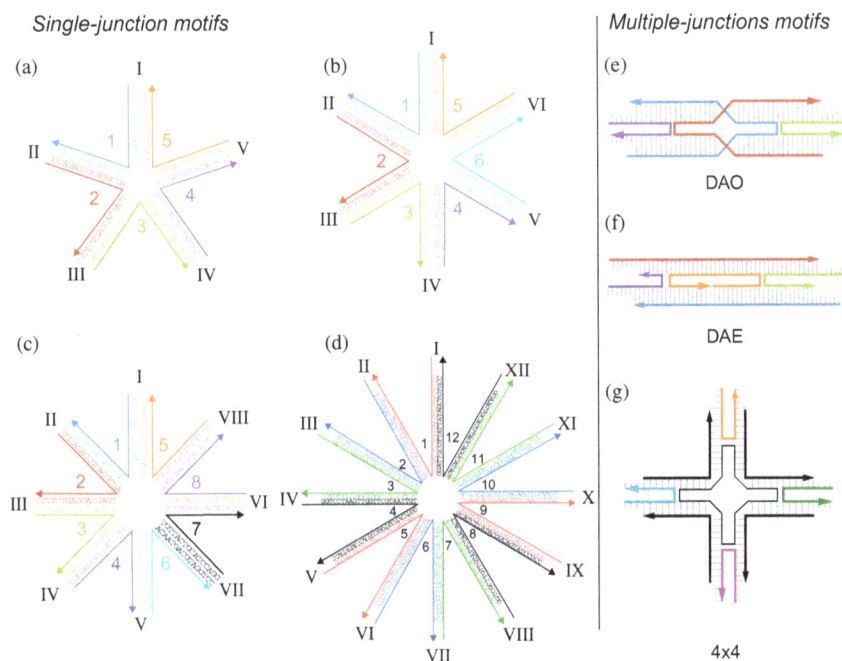

Figure 2.4. Branched DNA motifs containing one (a)–(d) or more (e)–(g) branch points. DNA junctions containing 5 (a), 6 (b), 8 (c) and even 12 (d) branches have been designed by SEQUIN and successfully obtained as single species. Note that increasing the number of branches does not necessitate redesigning the whole tile: additional arms are introduced into a pre-existent immobile junction in order to fulfill the required symmetry-minimization rules. More complex motifs realized with SEQUIN include the DX tile with an odd (DAO, e) or even (DAE, f) number of half-helical turns between the two crossovers and the more intricate 4 × 4 tile (g), formed by the self-assembly of nine oligonucleotides into four four-way junctions. Adapted with permission from Ref. [40].

Though being a very useful tool, SEQUIN still requires a significant amount of manual operation in the sequence optimization process and uncommon informatics skills. Therefore, alternative programs have been developed in the past few years, aiming at reducing the manual input from the user and attempting to provide user-friendly interfaces and graphical representations of the output structures. A significant example is the CANADA package [34], employed for the automated design of separate and concatenated oligonucleotides for the DNA-directed immobilization of proteins on a microarray [35, 36] and for the design of complex branched motifs [37]. Other relevant examples are GIDEON [38], Uniquimer 3D [39] and NanoEngineer-1 (available free of charge at http://www.nanoengineer-1.com). Those programs provide a virtual graphical model of the DNA motif and are supplemented with elementary algorithms for minimization of the energy of the construct [39]. Through an iterative relaxation process, the structural elements of the motif are fixed into a configuration of minimal strain, thus obtaining a rough estimate of the final most probable arrangement of helical segments in the target structure. In this way, it is possible for example to visualize the intrinsic curvature of a single tile, which must be taken into consideration when binding multiple tiles together, as explained in the next section.

2.3.2 *Tile-to-tile assembly*

The initial vision and final goal of structural DNA nanotechnology is to construct molecular architectures of desired geometrical features. Therefore, once single junctions are realized, the next step is to connect them into networks. This is achieved through sticky-ends cohesion, i.e. through hybridization of complementary single-stranded segments protruding from the 5' or 3' termini of junction's arms. Of course, also at this stage certain design criteria must be observed.

Two main parameters define the geometry of tile-to-tile assemblies: (i) the angles between the double-helical segments emerging from the crossovers and (ii) the distance between joined crossovers. The angles at the central junction of a tile are normally defined by the length of the single-stranded segments inserted between adjacent arms and from the total number of arms among which the mechanical stress is distributed. For example, inserting three thymine bases between two adjacent arms of a three-arms junction one obtains a motif with a C3 rotational symmetry, which tiles the plane into a regular hexagonal pattern (Figure 2.5(a)).

Such a tiling is also referred to as 6.6.6, indicating that at each junction three hexagons meet at their vertices tiling the plane into three identical angles of 120° [41]. By lowering the sequence symmetry of the system, distinct Archimedean tiling can be obtained [24]. Starting for example from the same three-arms junction, the number of thymine loops at the junction can be varied such to result into one angle of 90° and two

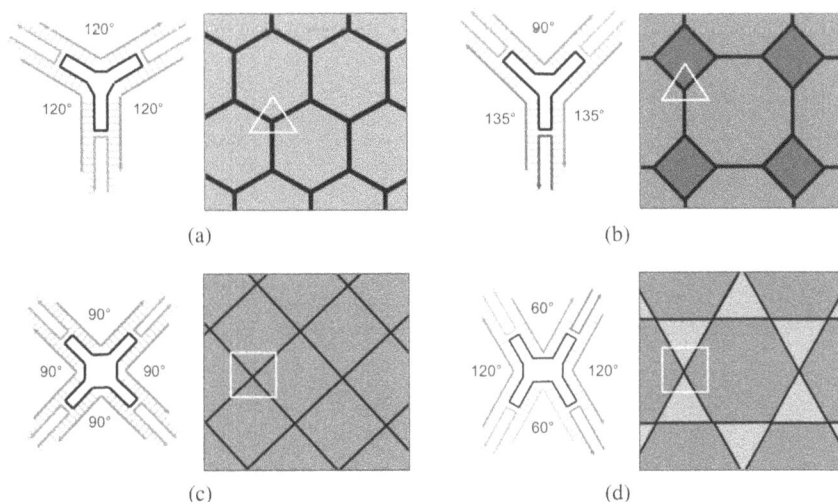

Figure 2.5. Platonic and Archimedean tilings of the plane. Using a sequence symmetric three-arms junction (with three T₃ loops), the rotational C3 symmetry leads to identical 120° angles between the tile arms, which in turn allows for the formation of platonic tilings of the kind 6.6.6 (a). Lowering the sequence symmetry of the tile (still keeping the same number of T loops), enables to tune the tile angles such to get 4.8.8 Archimedean patterns (b). In a similar way, a 4 × 4 tile of sequence and topological C4 symmetry (with four identical T₄ loops) results in the formation of 4.4.4.4 tessellations (c). An analogous tile of lower symmetry (two opposite T₃ and T₅ loops) (d) leads instead to 3.6.3.6 tilings of the plane.

angles of 135°. Such a motif will tile the plane into regular squares and octagons in a 4.8.8 pattern (Figure 2.5(b)). Similarly, using a 4 × 4 tile, regular square-like patterns can be achieved (4.4.4.4, Figure 2.5(c)) or 3.6.3.6 tessellations (Figure 2.5(d)).

Of course, one should consider that the torsional stress at the central junction imposed by the connection between adjacent arms gives rise to dihedral rather than plane angles and that this will finally end up into a precise tile curvature. Such an intrinsic curvature is of fundamental importance for the final architecture of the assembly product, which can assume completely different shapes depending on the second parameter of tile-to-tile association, i.e. the distance between adjacent crossovers. Considering n as the number of half-helical turns between joined crossovers, an even

(a)

(b)

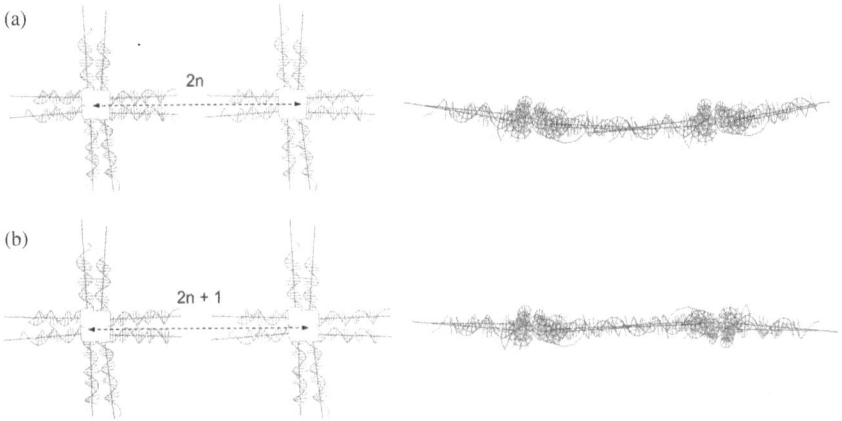

Figure 2.6. Tile-to-tile assembly. Graphical model of 4 × 4 tile extended lattices obtained by the Uniquimer 3D software. Visual models were obtained after an iterative process of energy minimization, which leads to the conformation of minimal strain. As the 4 × 4 tile is not a planar structure, sticky-ends cohesion between adjacent tiles leads to formation of large periodic structures, whose architecture depends on the number of half-helical turns between adjacent junctions. Whereas an even number of half-helical turns between interconnected junctions results in accumulation of tile curvatures and formation of tubular structures (a), an odd number of half-helical turns results in curvature compensation (also called corrugation), thus yielding to quasi-planar lattices (b). Adapted with permission from Ref. [40].

multiple of half-helical turns ($2n$, Figure 2.6(a)) will result in accumulation of the intrinsic curvature of the tile and formation of a closed assembly [42]. On the contrary, cohesion of adjacent tiles through an odd multiple of half-helical turns ($2n + 1$; Figure 2.6(b)) will cause them to face up and down alternately (according to a so-called "corrugation strategy") with formation of a quasi-planar array [33].

Finally, the thermal stability of tile-to-tile interaction is directly related to the strength of sticky-ends cohesion. This can be typically enhanced by increasing the CG ratio and/or the length of the complementary DNA segments. On the other hand, excessive hybridization forces should be avoided, as this might result in the trapping of misfolded structures thus reducing formation of the desired network. As a rule of thumb, three to five nucleobases long sticky-ends are normally sufficient to ensure specificity and stability of the newly formed double-helical "bond," still enabling self-correction mechanisms to take place.

2.3.3 *Platonic and Archimedean tilings*

Tilings or tessellations are ways of filling up the plane with tiles [43]. Normally, tilings satisfy the following conditions: (i) the tiles are regular polygons, (ii) the tiling is edge-to-edge: this means that two tiles intersect along a common edge, only at a common vertex or not at all; (iii) all the vertices are of the same type: this means that the same types of polygons meet in the same order (ignoring orientation) at each vertex. There are only three tilings that use only one type of tiles. These are called *Platonic* or regular tilings. On the other hand, there are eight tilings that use more than one type of tiles. They are called *Archimedean* or semi-regular tilings [44]. To indicate such geometric patterns, the vertex configuration is given. This is a sequence of numbers representing the number of sides of the faces going around the vertex. For example, the notation 4.8.8 describes a vertex that has three faces around it, namely a square and two octagons. Several theoretical frameworks have been developed for predicting the outcome of self-assembly processes leading to tiling of the plane [45]. In the field of DNA self-assembly, Yan and coworkers recently described a design strategy that enables formation of Archimedean tilings by using only one type of tile with lower symmetry (Figure 2.2(b)), thus enlarging the toolbox of DNA tile-based self-assembly and expanding the complexity of structures attainable by DNA-based tessellation [24].

2.3.4 *Single-stranded tiles*

Tile-based assembly normally leads to formation of periodic structures of unlimited size. In 2012, Yin and coworkers developed a design approach to construct finite yet complex shapes from a large number (hundreds to even one thousand) of uniquely addressable tiles [25]. The method relies on the simplest form of tile: a single-stranded DNA (SST) constituted by 42 bases divided into four domains, each binding to four local neighboring tiles during self-assembly (Figure 2.2(c)). Considering each SST as a pixel, the method is based on the generation of a master strand collection that corresponds to a 310-pixel canvas, from which appropriate subsets of strands are selected to achieve a desired 2D shape. In this way, SST assembly provides a simple and modular framework for constructing aperiodic nanostructures of complex addressability. The same concept has been extended to 3D shapes, using so-called "DNA-bricks" as building blocks of 3D canvas (Figure 2.2(d)) [26]. Each brick is constituted by 32-nuleotide long strands, which bind to four neighboring modular components through 8-bp interactions. By selecting subsets of bricks from the molecular canvas, the authors constructed a series of distinct 3D shapes with sophisticated surface features as well as intricate interior cavities and tunnels.

2.4 Scaffold-based Designs

An extraordinary breakthrough in the construction of nanometer sized DNA objects occurred in 2006 with the introduction of the scaffolded DNA-origami method by Rothemund [19] (for reviews on this subject, see Ref. [46]). Similarly to the Japanese art of paper folding, the DNA-origami technique folds a long single-stranded DNA scaffold into a desired shape by the help of hundreds of short oligonucleotides, called staple strands. The arrangement of crossovers between the connected helices of a DNA origami defines the shape of the structure. This can be planar, with helical axes arranged in an antiparallel fashion along single monolayers, or space-filled, with helical axes oriented one another into a hexagonal or square pattern (Figure 2.7).

2.4.1 *Monolayer DNA origami*

In the original work of Rothemund [19], several planar origami structures were produced, ranging from simple rectangular shapes to more complex forms, such as stars, triangles, as well as non-geometric figures as for example smiley faces (Figure 2.7(a), left side). These objects were generated by following a common design rule: every helix within the structure is connected to two neighboring helices by a regular pattern of crossovers interspaced by 1.5 helical turns, which for a B-type DNA, correspond to about 16 base pairs. This register of crossovers generates interhelical connections every 180° thus leading to a single layer of helices arranged into a planar sheet.

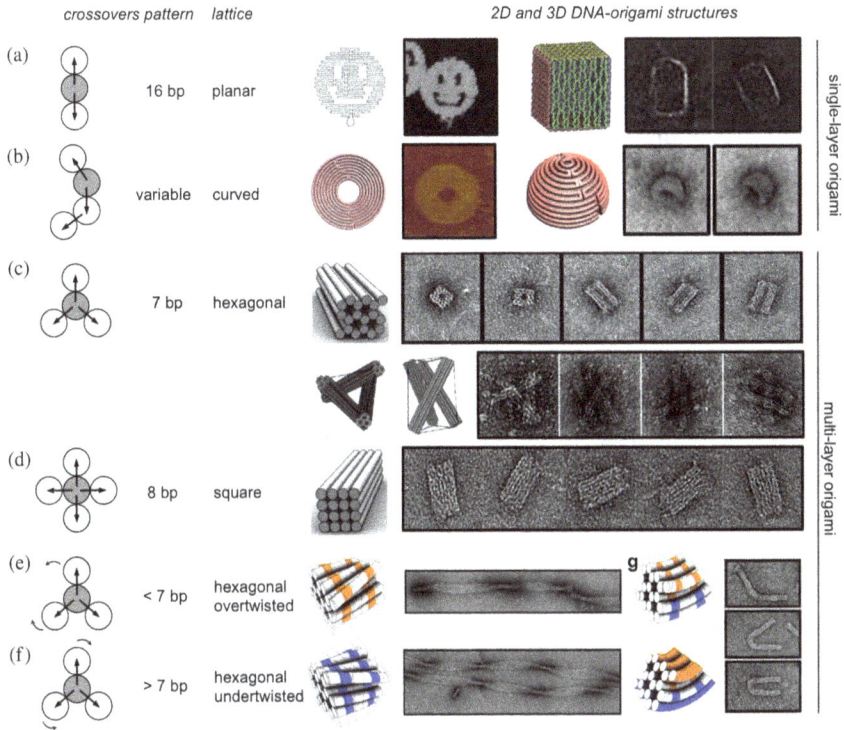

Figure 2.7. Schematic representation of crossover patterns and thus-generated lattices used for the construction of 2D and 3D DNA-origami structures. DNA helices are indicated by circles and are viewed along their central axis. Interhelical crossovers between a reference helix (grey circle) and its neighboring helices (white circles) are indicated as black arrows. The number of base pairs between consecutive crossovers along the same helical axis is given, as well as the resulting 2D or 3D arrangement of the helices. Single-layer (a) and (b) and multi-layer (c)–(g) DNA-origami structures can be generated by suitable engineering of crossovers, thus leading to a large variety of shapes, including vacant, filled, twisted and curved objects. Adapted with permission from Ref. [22].

The same design principle can also be used to generate 3D structures. To this end, an origami is assembled which contains individual 2D sheets. Those sheets are then connected at specific angles by an additional set of crossovers between interfacing helices at the edges (Figure 2.7(a), right side). This generates 3D objects with an internal cavity, such as

prisms-like structures with three, four or six faces [47], or closed polyhe-
dra, such as a tetrahedron [48] or cubes [49, 50].

Monolayer origami objects with complex curvatures were also produced by bending DNA helices along their central axis and linking them
together by a network of latitudinal and longitudinal crossovers to produce 2D and 3D structures, respectively (Figure 2.7(b)) [51].

Contrary to previously reported methods, this network of crossovers
does not strictly obey the rule of 10.5 bp/turn, but rather permits a certain
degree of structural flexibility (from 9 to 11 bp/turn), which allows for a
more accurate tuning of the DNA curvature. This in turns gives access to
shapes such as planar concentric rings, spheres and hemispheres, which
are not reachable by conventional origami methods.

2.4.2 *Space-filled DNA origami*

One major limitation of single-layer DNA origami is their relatively weak
resistance to mechanical stress. To address this problem, more rigid 3D
DNA objects have been developed by either packing multiple helices into
a space-filling structure [52] or taking advantage of tensegrity rules
(Figure 2.7(c)) [53]. Multi-layer DNA-origami structures are densely
packed arrays of antiparallel helices interconnected through a defined 3D
arrangement of crossovers. The register of such crossovers, that is their
relative angular displacement with respect to the axis of the helix, defines
the way adjacent helices are interconnected and therefore determines the
geometry of the basic building block. The first example of multi-layer
DNA origami was reported by Shih and coworkers [52]. In their design
strategy, every helix is connected to three adjacent helices by crossovers
relatively oriented at 120°, spaced by 7 base pairs along the helical axis.
The resulting superstructure has therefore a hexagonal cross section similar to a honeycomb lattice (Figure 2.7(c), upper panel). The generality of
the method was demonstrated by successful formation of a series of 3D
shapes resembling for example a monolith or a square nut. Following the
same design principle, DNA objects were built with a rectangular [54] or
closed-packed hexagonal cross-section [55] enabling construction of flat-
ter surfaces, smaller cavities and denser structures (Figure 2.7(d)). Shortly
later, Dietz *et al.* were able to engineer supertwisted or bended structures

through deletion or addition of a distinct number of base pairs within helical arrays (Figures 2.7(e)–2.7(g)) [56].

In a different approach, Liedl and coworkers introduced mechanical stability into 3D DNA objects taking advantage of the tensegrity concept (Figure 2.7(c) lower panel) [53]. The term "tensegrity" stays for "tensional integrity" and indicates geometrical integrity upon tension. This architectural principle is often applied in macroscopic buildings, where geometric objects comprised of stiff sticks ("struts") are connected by flexible linkers ("tendons"). Since struts push outward and tendons pull inward, the balance between the two forces leads to stable and mechanically rigid structures. As demonstrated by Liedl *et al.* [53], 3D tensegrity prisms can be constructed which are composed of rigid DNA bundles of three to six helices, working as compressive-resistant struts, held in place by ssDNA segments, which function as entropic spring tendons.

It should be stressed that such a rapid development of the DNA-origami technology and the realization of sophisticated structures would have not been equally effective without the help of adequate design programs. Fortunately, software solutions for designing 2D [57] and 3D [58] origami are in continuous evolution and made available on a public domain. Semi-automated design tools such as caDNAno (http://cadnano .org) enormously reduce the time required for generating a new structure and, more important, drastically lower the probability of human errors which may occur in a completely manual process. In addition, programs such as CanDo (http://cando-dna-origami.org) are available for the analysis of the designed structures enabling to predict for example their thermal fluctuations and geometrical properties. These programs are also constantly upgraded with new features and functions, which make possible to apply variations on design strategies in an easier way [11], thus notably boosting the potential of the technology.

Finally, alternative strategies have emerged that seek to scale up the dimensions of DNA nanostructures thus trying to fill up the gap between the nano- and macroscopic world in a rationally designed manner. Examples of such strategies include the use of so-called "short scaffold parity strands" [59] or a set of different nucleobase-specific forces, shape-complementarity and template-related effects to connect origami shapes together into large hierarchical assemblies [7].

2.4.3 Wireframe DNA origami

In the original DNA-origami design, the scaffold strand and corresponding double helices are organized into parallel, raster-fill patterns. This results into tightly packed structures, constituted by domains of adjacent antiparallel helices joined by DX motifs. Few years later, Yan and coworkers firstly faced the challenge to construct scaffold-based structures of wireframe geometry (Figure 2.8).

Wireframe structures are realized by linking multi-arm junction units together, each one containing 2 to 12 arms, thus allowing for the non-parallel alignment of DNA helices. This results in the formation of 2D net-like surfaces with various sizes and shaped cavities as well as 3D polyhedrons, curved solids and multi-layered frameworks.

Using a four-arms junction as the basic repetitive unit, Yan and coworkers succeeded in folding the scaffold into planar gridiron architectures (Figure 2.8(a)) [60]. Experimental observations revealed the formation of rhomboid rather than square structures, due to the flexibility of the motif and its relaxed conformation into a right-handed twist with a 60° torsion angle. The strategy was extended to the third dimension either by controlling the curvature of the single motifs to get distorted shapes (Figure 2.8(b)), or by stacking or intertwining multiple layers of 2D gridiron lattices at selected connection points (Figure 2.8(c)).

Multi-layered DNA frameworks with well-controlled geometry, sustained mechanical stability and tunable cavity size have been lately created by a novel layered-crossover motif (named as LX) [61]. The LX motif is based on the DX motif with the difference that the DNA helices connected by the crossovers are not belonging to the same plane, but to two parallel layers (Figure 2.8(d)). In this way, DNA helices can be aligned in a non-parallel arrangement through stacks of layers, thus enriching the toolbox of structural DNA nanotechnology (Figure 2.8(e)). The same group demonstrated the versatility of the idea by implementing arbitrarily designed connections between selected multi-arms junctions in a 2D and 3D space (Figure 2.8(f)) [28].

The design relies on a four-step process. In the first step, all connections between the vertices of the wireframe structure are converted into double lines. In the second step, lines are looped and bridged such to

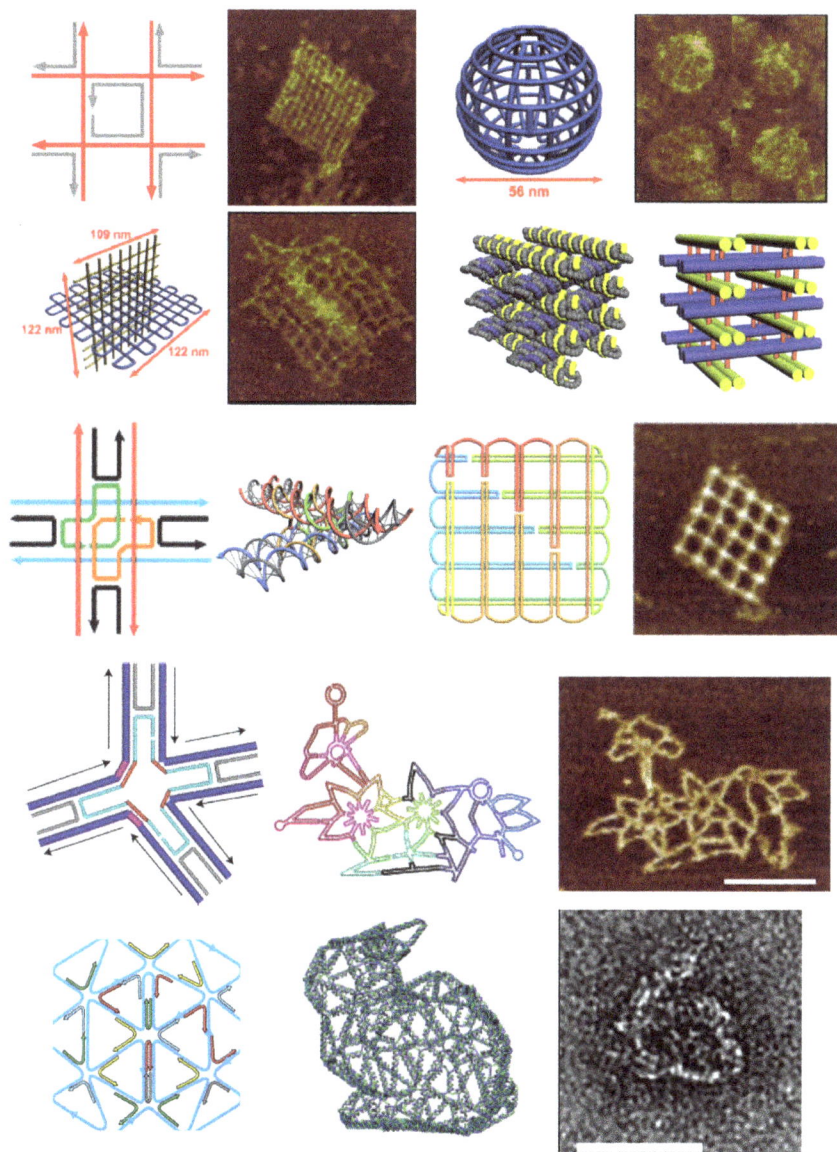

Figure 2.8.   Wireframe-based DNA origami designs. Using a Holliday junction as building unit, 2D (a), curved (b) and lattice like (c) gridiron structures can be obtained [60]. Starting instead from DX motifs spanning two adjacent layers (LX motifs, d), rigid multi-layers frameworks can be produced (e) [61]. The same principle applied to multi-arms junctions with arbitrarily designed connections allows constructing sophisticated 2D and 3D patterns (f) [28]. A latest strategy instead, based on the graph theory, folds the scaffold according to a triangulation pattern, eventually leading to formation of polyhedral meshes (g) [27]. Adapted with permission from Refs. [27, 28, 60, 61].

ensure the travelling of the scaffold strand through all vertices only once, enabling also a circular path. In a third step, the complementary staple strands are added to the scaffold lines to generate DX tiles spaced by a full number of helical turns. This is done to minimize strain and allow the scaffold to follow the predesigned path. Final step is to adjust the angles at the vertices. This is achieved by inserting a T_n loop of appropriate length into the staple strands surrounding the vertex and leaving a certain number of nucleobases in the scaffold — opposite to the T_n loop — as unpaired. In this way, each vertex is provided with a certain degree of structural flexibility such that the arms are allowed to bend and position in the most favorable manner, eventually generating the desired set of angles. The method has been applied for the construction of simple platonic tilings, curved patterns as well as intricate and arbitrary shapes in the form of three flowers and a bird (Figure 2.8(f)). In addition, 3D architectures of Archimedean symmetry have been also produced using Schlegel diagrams [62] to generate the looping path of the scaffold.

Similar polyhedral meshes were obtained by Högberg and coworkers using a completely different design approach [27], based on the "Chinese postman tour" [63] problem in graph theory (Figure 2.8(g)). This design strategy relies on three main principles: (i) the meshes of the structure are triangulated in order to optimize structural rigidity; (ii) the edges of the structure are constituted by single double-helical segments joined one another at common vertices and (iii) these latter should not cross, that is, the scaffold should pass only once through all vertices. Also in this case, the overall design process is split into four steps, which can be run with the help of a software tool specifically created by the authors (vHelix, available free of charge at http://www.vhelix.net). In a first step, the 3D polygonal mesh is created using the 3D graphical interface. In a second step, the routing of the scaffold through all the edges of the mesh is generated. Third, the least strained DNA helix arrangement is determined and finally, fine-tuning of the design through a "relaxation algorithm" and assignment of the staple strands is performed. In this way, different polygonal 3D meshes were generated, including an icosahedron, a nicked torus, a bottle and a version of the Stanford bunny.

2.5 DNA Aptamers: From Basics to Applications

Aptamers (from the Latin *aptus* — fit, and Greek *meros* — part) are defined as DNA or RNA molecules — typically <100-mer, that bind to a specific target molecule of bioanalytical significance with high affinity and specificity. Aptamers are evolved by an evolutionary process, defined as systematic evolution of ligands by exponential enrichment (SELEX) in which sequences with a certain conformation capable of binding to the target of interest emerge and dominate the oligonucleotides pool. The SELEX strategy was firstly described in 1990 by the laboratories of Gold and Szostak [64, 65]. The Gold lab used the term SELEX for their process of selecting RNA ligands against T4 DNA polymerase. On the other hand, the Szostak lab defined the term "*in vitro* selection," selecting RNA ligands against various organic dyes. The Szostak lab also coined the term "aptamer." Two years later, the Szostak lab and Gilead Sciences, employed *in vitro* selection schemes to evolve single-stranded DNA ligands for organic dyes and human coagulant, thrombin. From that time, the researchers started to employ DNA aptamers more favorably than RNA aptamers, given the intrinsic greater chemical stability of DNA in comparison with RNA.

SELEX consists of basic steps including incubation of targets with oligonucleotide library (typically 10^{13}–10^{16} random sequences), isolation of the oligonucleotide — target complexes from unbound sequences and finally amplification of the bound sequences by PCR in order to obtain an enriched pool for the next round of selection (see Figure 2.9). By replicating the selection process, it is possible to enrich the bound sequences.

Throughout the entire process, the most critical step is the separation of the bound sequences from unbound sequences, especially for SELEX using purified targets. Thus many modified SELEX strategies have been proposed to simplify this step or to enhance the efficiency of separation [66]. The selected DNA or RNA sequences are finally cloned into bacteria and sequenced with the aim to obtain a pool of optimal individual aptamer candidates, which are finally labeled with reporters.

Aptamers are known also as nucleic acid antibodies and have unique features, such as ease of chemical synthesis, excellent chemical stability in physiological conditions, low molecular weight, lack of

W. Pfeifer et al.

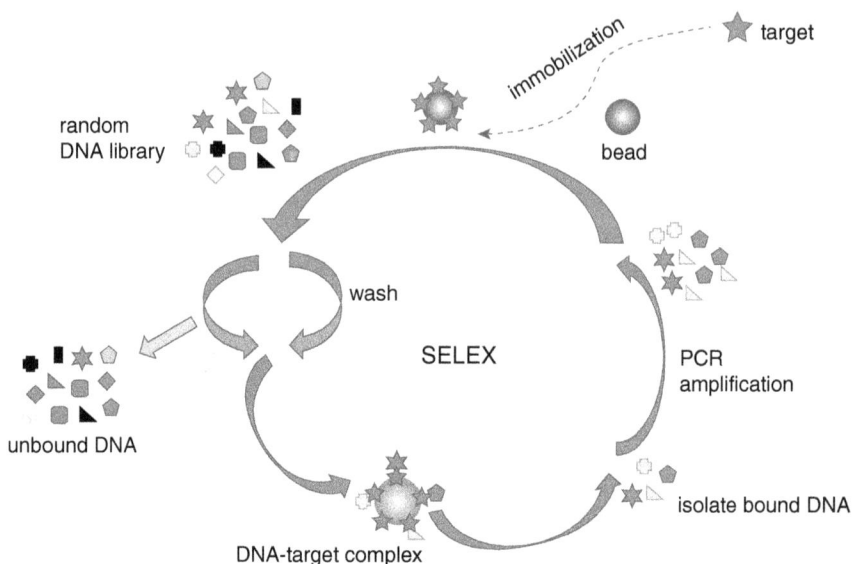

Figure 2.9. Scheme of systematic evolution of ligands by exponential enrichment (SELEX) process for the synthesis of an aptamer for a protein target. Reproduced with kind permission from Ref. [67].

immunogenicity, and ease of modification in comparison with protein counterparts. They have the ability to bind to different targets such as metal ions, small molecules, proteins, whole live cells, viruses and bacteria with high affinity and specificity. Such features make aptamers very promising candidates as molecular probes for novel class of electrochemical biosensors, as active elements for the recognition of extracellular matrix signatures of cancer cells and target-specific ligands for therapeutic purposes.

Given the possibility to engineer nucleic acid bases, a large number of aptamer-based sensing systems have been developed for the efficient detection of a wide range of molecules. Just to make some examples, Plaxco *et al.* designed a series of aptamer-based electrochemical sensors to detect ATP, cocaine and so on [68]; Zhao *et al.* designed a fluorescent biosensor to detect proteins [69]; Ricci *et al.* designed allosterically controllable, metal–ion triggered molecular switches to specifically recognize mercury(II) or silver(I) ions [70]. Remarkably, aptamers can be generated also to recognize target cells, including those expressing specific proteins of interest. In this case, the selection process is defined cell-SELEX [69]. Typically, these aptamers are selected against live or — less commonly — fixed cells. Partitioning between unbound sequences and cell-bound sequences is carried out by centrifugation or washing. Remarkably, at least one type of control cell has to be used for counter-selection in order to get rid of sequences that bind to proteins present at surface of both cell types.

One of the major issues of initial SELEX technique is its time-consuming, labor-intensive nature. This problem was solved by the lab of C. Cox that was able to automatize *in vitro* selection process, so reducing the duration of the selection to some days [71, 72]. In addition, another significant issue of DNA aptamers is that most of the reported synthesized aptamers are selected and applied for *in vitro* studies, whereas the *in vivo* stability of these aptamers is still not perfectly known at present. Unmodified nucleic acid probes are susceptible to nuclease digestion, so they are unstable in both intercellular and intra-cellular environments. Therefore, innovation in SELEX protocols will take into consideration aptamers with higher resistance under clinical applications [66].

References

1. Seeman, N.C. (1982). Nucleic acid junctions and lattices. *J. Theor. Biol.* **99**(2), 237–247.
2. Frutos, A.G., Liu, Q., Thiel, A.J., Sanner, A.M., Condon, A.E., Smith, L.M., and Corn, R.M. (1997). Demonstration of a word design strategy for DNA computing on surfaces. *Nucleic Acids Res.* **25**(23), 4748–4757.
3. Liu, Q., Wang, L., Frutos, A.G., Condon, A.E., Corn, R.M., and Smith, L.M. (2000). DNA computing on surfaces. *Nature.* **403**(6766), 175–179.
4. Feldkamp, U., Rauhe, H., and Banzhaf, W. (2003). Software tools of DNA sequence design. *Genet. Program. Evol. M.* **4**(2), 153–171.
5. Seeman, N.C., and Kallenbach, N.R. (1983). Design of immobile nucleic acid junctions. *Biophys. J.* **44**(2), 201–209.
6. Seeman, N.C., and Kallenbach, N.R. (1994). DNA branched junctions. *Ann. Rev. Biophys. Biomol. Struct.* **23**, 53–86.

7. Pfeifer, W., and Saccà, B. (2016). From nano to macro through hierarchical self-assembly: The DNA paradigm. *Chembiochem.* **17**, 1063–1080.

8. Whitelam, S. (2015). Hierarchical assembly may be a way to make large information-rich structures. *Soft Mat.* **11**(42), 8225–8235.

9. Frenkel, D. (2015). Order through entropy. *Nat. Mater.* **14**(1), 9–12.

10. Gerling, T., Wagenbauer, K.F., Neuner, A.M., and Dietz, H. (2015). Dynamic DNA devices and assemblies formed by shape-complementary, non-base pairing 3D components. *Science.* **347**(6229), 1446–1452.

11. Woo, S., and Rothemund, P.W. (2011). Programmable molecular recognition based on the geometry of DNA nanostructures. *Nat. Chem.* **3**(8), 620–627.

12. Rajendran, A., Endo, M., Katsuda, Y., Hidaka, K., and Sugiyama, H. (2011). Programmed two-dimensional self-assembly of multiple DNA origami jigsaw pieces. *ACS Nano.* **5**(1), 665–671.

13. Suzuki, Y., Endo, M., and Sugiyama, H. (2015). Lipid-bilayer-assisted two-dimensional self-assembly of DNA origami nanostructures. *Nat. Commun.* **6**, 8052.

14. Kocabey, S., Kempter, S., List, J., Xing, Y., Bae, W., Schiffels, D., Shih, W.M., Simmel, F.C., and Liedl, T. (2015). Membrane-assisted growth of DNA origami nanostructure arrays. *ACS Nano.* **9**(4), 3530–3539.

15. Woo, S., and Rothemund, P.W. (2014). Self-assembly of two-dimensional DNA origami lattices using cation-controlled surface diffusion. *Nat. Commun.* **5**, 4889.

16. Aghebat Rafat, A., Pirzer, T., Scheible, M.B., Kostina, A., and Simmel, F.C. (2014). Surface-assisted large-scale ordering of DNA origami tiles. *Angew. Chem. Int. Ed. Engl.* **53**(29), 7665–7668.

17. Sun, X., Hyeon Ko, S., Zhang, C., Ribbe, A.E., and Mao, C. (2009). Surface-mediated DNA self-assembly. *J. Am. Chem. Soc.* **131**(37), 13248–13249.

18. Schulman, R., and Winfree, E. (2007). Synthesis of crystals with a programmable kinetic barrier to nucleation. *Proc. Natl. Acad. Sci. USA.* **104**(39), 15236–15241.

19. Rothemund, P.W. (2006). Folding DNA to create nanoscale shapes and patterns. *Nature.* **440**(7082), 297–302.

20. Yan, H., LaBean, T.H., Feng, L., and Reif, J.H. (2003). Directed nucleation assembly of DNA tile complexes for barcode-patterned lattices. *Proc. Natl. Acad. Sci. USA.* **100**(14), 8103–8108.

21. Shih, W.M., Quispe, J.D., and Joyce, G.F. (2004). A 1.7-kilobase single-stranded DNA that folds into a nanoscale octahedron. *Nature.* **427**(6975), 618–621.

22. Sacca, B., and Niemeyer, C.M. (2012). DNA origami: The art of folding DNA. *Angew. Chem. Int. Ed. Engl.* **51**(1), 58–66.

23. Hansen, M.N., Zhang, A.M., Rangnekar, A., Bompiani, K.M., Carter, J.D., Gothelf, K.V., and LaBean, T.H. (2010). Weave tile architecture construction strategy for DNA nanotechnology. *J. Am. Chem. Soc.* **132**(41), 14481–14486.

24. Zhang, F., Jiang, S., Li, W., Hunt, A., Liu, Y., and Yan, H. (2016). Self-assembly of complex DNA tessellations by using low-symmetry multi-arm DNA tiles. *Angew. Chem. Int. Ed. Engl.* **55**(31), 8860–8863.

25. Wei, B., Dai, M., and Yin, P. (2012). Complex shapes self-assembled from single-stranded DNA tiles. *Nature.* **485**(7400), 623–626.

26. Ke, Y., Ong, L.L., Shih, W.M., and Yin, P. (2012). Three-dimensional structures self-assembled from DNA bricks. *Science.* **338**(6111), 1177–1183.

27. Benson, E., Mohammed, A., Gardell, J., Masich, S., Czeizler, E., Orponen, P., and Hogberg, B. (2015). DNA rendering of polyhedral meshes at the nanoscale. *Nature.* **523**(7561), 441–444.

28. Zhang, F., Jiang, S., Wu, S., Li, Y., Mao, C., Liu, Y., and Yan, H. (2015). Complex wireframe DNA origami nanostructures with multi-arm junction vertices. *Nat. Nanotechnol.* **10**(9), 779–784.

29. Borer, P.N., Dengler, B., Tinoco, I., Jr., and Uhlenbeck, O.C. (1974). Stability of ribonucleic acid double-stranded helices. *J. Mol. Biol.* **86**(4), 843–853.

30. Wang, Y.L., Mueller, J.E., Kemper, B., and Seeman, N.C. (1991). Assembly and characterization of five-arm and six-arm DNA branched junctions. *Biochemistry.* **30**(23), 5667–5674.

31. Wang, X., and Seeman, N.C. (2007). Assembly and characterization of 8-arm and 12-arm DNA branched junctions. *J. Am. Chem. Soc.* **129**(26), 8169–8176.

32. Fu, T.J., and Seeman, N.C. (1993). DNA double-crossover molecules. *Biochemistry.* **32**(13), 3211–3220.

33. Yan, H., Park, S.H., Finkelstein, G., Reif, J.H., and LaBean, T.H. (2003). DNA-templated self-assembly of protein arrays and highly conductive nanowires. *Science.* **301**(5641), 1882–1884.

34. Feldkamp, U. (2009). CANADA: Designing nucleic acid sequences for nanobiotechnology applications. *J. Comput. Chem.* **31**(3), 660–663.

35. Feldkamp, U., Schroeder, H., and Niemeyer, C.M. (2006). Design and evaluation of single-stranded DNA carrier molecules for DNA-directed assembly. *J. Biomol. Struct. Dyn.* **23**(6), 657–666.

36. Feldkamp, U., Wacker, R., Schroeder, H., Banzhaf, W., and Niemeyer, C.M. (2004). Microarray-based *in vitro* evaluation of DNA oligomer libraries designed in silico. *Chemphyschem.* **5**(3), 367–372.

37. Sacca, B., Meyer, R., Feldkamp, U., Schroeder, H., and Niemeyer, C.M. (2008). High-throughput, real-time monitoring of the self-assembly of DNA nanostructures by FRET spectroscopy. *Angew. Chem. Int. Ed. Engl.* **47**(11), 2135–2137.

38. Birac, J.J., Sherman, W.B., Kopatsch, J., Constantinou, P.E., and Seeman, N.C. (2006). Architecture with GIDEON, a program for design in structural DNA nanotechnology. *J. Mol. Graph. Model.* **25**(4), 470–480.

39. Zhu, J., Wei, B., Yuan, Y., and Mi, Y. (2009). UNIQUIMER 3D, a software system for structural DNA nanotechnology design, analysis and evaluation. *Nucleic Acids Res.* **37**(7), 2164–2175.

40. Saccà, B., Sprengel, A., and Feldkamp, U. (2013). *De novo* design of nucleic acid structures. In *De novo Molecular Design*, Wiley-VCH Verlag GmbH & Co. KGaA, Weinheim, Germany. pp. 495–517.

41. He, Y., Chen, Y., Liu, H., Ribbe, A.E., and Mao, C. (2005). Self-assembly of hexagonal DNA two-dimensional (2D) arrays. *J. Am. Chem. Soc,* **127**(35), 12202–12203.

42. Liu, D., Park, S.H., Reif, J.H., and LaBean, T.H. (2004). DNA nanotubes self-assembled from triple-crossover tiles as templates for conductive nanowires. *Proc Natl. Acad. Sci. USA.* **101**(3), 717–722.

43. Torquato, S., and Jiao, Y. (2009). Dense packings of the Platonic and Archimedean solids. *Nature.* **460**(7257), 876–879.

44. Millan, J.A., Ortiz, D., van Anders, G., and Glotzer, S.C. (2014). Self-assembly of Archimedean tilings with enthalpically and entropically patchy polygons. *ACS Nano.* **8**(3), 2918–2928.

45. Mannige, R.V., and Whitelam, S. (2016). Predicting the outcome of the growth of binary solids far from equilibrium. *Phys. Rev. E.* **93**, 042136.

46. Torring, T., Voigt, N.V., Nangreave, J., Yan, H., and Gothelf, K.V. (2011). DNA origami: A quantum leap for self-assembly of complex structures. *Chem. Soc. Rev.* **40**, 5636–5646.

47. Endo, M., Hidaka, K., Kato, T., Namba, K., and Sugiyama, H. (2009). DNA prism structures constructed by folding of multiple rectangular arms. *J. Am. Chem. Soc.* **131**(43), 15570–15571.

48. Ke, Y., Sharma, J., Liu, M., Jahn, K., Liu, Y., and Yan, H. (2009). Scaffolded DNA origami of a DNA tetrahedron molecular container. *Nano. Lett.* **9**(6), 2445–2447.

49. Andersen, E.S., Dong, M., Nielsen, M.M., Jahn, K., Subramani, R., Mamdouh, W., Golas, M.M., Sander, B., Stark, H., Oliveira, C.L., Pedersen, J.S., Birkedal, V., Besenbacher, F., Gothelf, K.V., and Kjems, J. (2009). Self-assembly of a nanoscale DNA box with a controllable lid. *Nature*. **459**(7243), 73–76.

50. Kuzuya, A., and Komiyama, M. (2009). Design and construction of a box-shaped 3D-DNA origami. *Chem. Commun.* (28), 4182–4184.

51. Han, D., Pal, S., Nangreave, J., Deng, Z., Liu, Y., and Yan, H. (2011). DNA origami with complex curvatures in three-dimensional space. *Science*. **332**(6027), 342–346.

52. Douglas, S.M., Dietz, H., Liedl, T., Hogberg, B., Graf, F., and Shih, W.M. (2009). Self-assembly of DNA into nanoscale three-dimensional shapes. *Nature*. **459**(7245), 414–418.

53. Liedl, T., Hogberg, B., Tytell, J., Ingber, D.E., and Shih, W.M. (2010). Self-assembly of three-dimensional prestressed tensegrity structures from DNA. *Nat. Nanotechnol.* **5**(7), 520–524.

54. Ke, Y., Douglas, S.M., Liu, M., Sharma, J., Cheng, A., Leung, A., Liu, Y., Shih, W.M., and Yan, H. (2009). Multilayer DNA origami packed on a square lattice. *J. Am. Chem. Soc.* **131**(43), 15903–15908.

55. Ke, Y., Voigt, N.V., Gothelf, K.V., and Shih, W.M. (2012). Multilayer DNA origami packed on hexagonal and hybrid lattices. *J. Am. Chem. Soc.* **134**(3), 1770–1774.

56. Dietz, H., Douglas, S.M., and Shih, W.M. (2009). Folding DNA into twisted and curved nanoscale shapes. *Science*. **325**(5941), 725–730.

57. Andersen, E.S., Dong, M., Nielsen, M.M., Jahn, K., Lind-Thomsen, A., Mamdouh, W., Gothelf, K.V., Besenbacher, F., and Kjems, J. (2008). DNA origami design of dolphin-shaped structures with flexible tails. *ACS Nano*. **2**(6), 1213–1218.

58. Douglas, S.M., Marblestone, A.H., Teerapittayanon, S., Vazquez, A., Church, G.M., and Shih, W.M. (2009) Rapid prototyping of 3D DNA-origami shapes with caDNAno. *Nucleic Acids Res*. **37**(15), 5001–5006.

59. Ke, Y., Bellot, G., Voigt, N.V., Fradkov, E., and Shih, W.M. (2012). Two design strategies for enhancement of multilayer-DNA-origami folding: underwinding for specific intercalator rescue and staple-break positioning. *Chem. Sci.* **3**(8), 2587–2597.

60. Han, D., Pal, S., Yang, Y., Jiang, S., Nangreave, J., Liu, Y., and Yan, H. (2013). DNA gridiron nanostructures based on four-arm junctions. *Science*. **339**(6126), 1412–1415.

61. Hong, F., Jiang, S., Wang, T., Liu, Y., and Yan, H. (2016). 3D framework DNA origami with layered crossovers. *Angew. Chem. Int. Ed. Engl.* **55**(41), 12832–12835.

62. Sommerville, D.M.Y. (1929). *Introduction to the Geometry of N Dimensions.* E.P. Dutton. New York, E.P. Dutton and company.

63. Edmonds, J., and Johnson, E.L. (1973). Matching, Euler tours and the Chinese postman. *Math. Program.* **5**, 88–124.

64. Ellington, A.D., and Szostak, J.W. (1990). *In vitro* selection of RNA molecules that bind specific ligands. *Nature.* **346**(6287), 818–822.

65. Tuerk, C., and Gold, L. (1990). Systematic evolution of ligands by exponential enrichment: RNA ligands to bacteriophage T4 DNA polymerase. *Science.* **249**(4968), 505–510.

66. Lyu, Y., Chen, G., Shangguan, D., Zhang, L., Wan, S., Wu, Y., Zhang, H., Duan, L., Liu, C., You, M., Wang, J., and Tan, W. (2016). Generating cell targeting aptamers for nanotheranostics using Cell-SELEX. *Theranostics.* **6**(9), 1440–1452.

67. Meng, H.M., Liu, H., Kuai, H., Peng, R., Mo, L., and Zhang, X.B. (2016). Aptamer-integrated DNA nanostructures for biosensing, bioimaging and cancer therapy. *Chem. Soc. Rev.* **45**(9), 2583–2602.

68. Lubin, A.A., and Plaxco, K.W. (2010). Folding-based electrochemical biosensors: the case for responsive nucleic acid architectures. *Acc. Chem. Res.* **43**(4), 496–505.

69. Huang, Y., Chen, J., Zhao, S., Shi, M., Chen, Z.F., and Liang, H. (2013). Label-free colorimetric aptasensor based on nicking enzyme assisted signal amplification and DNAzyme amplification for highly sensitive detection of protein. *Anal. Chem.* **85**(9), 4423–4430.

70. Porchetta, A., Vallee-Belisle, A., Plaxco, K.W., and Ricci, F. (2013). Allosterically tunable, DNA-based switches triggered by heavy metals. *J. Am. Chem. Soc.* **135**(36), 13238–13241.

71. Cox, J.C., and Ellington, A.D. (2001). Automated selection of anti-protein aptamers. *Bioorg. Med. Chem.* **9**(10), 2525–2531.

72. Cox, J.C., Rajendran, M., Riedel, T., Davidson, E.A., Sooter, L.J., Bayer, T.S., Schmitz-Brown, M., and Ellington, A.D. (2002). Automated acquisition of aptamer sequences. *Comb. Chem. High. T. Scr.* **5**(4), 289–299.

Chapter 3

DNA Sensors for the Detection of Biomolecules and Biochemical Conditions

Birgitta R. Knudsen*, Anni H. Andersen*,
Magnus Stougaard†, Giuseppe Arrabito‡, Raffaella Suriano§
and Yi-Ping Ho¶,‖

*Department of Molecular Biology and Genetics
Aarhus University, Denmark
†Department of Clinical Medicine
Aarhus University, Denmark
‡Department of Physics and Chemistry
University of Palermo, Viale delle Scienze, Parco d'Orleans II
90128 Palermo, Italy
§Department of Chemistry, Materials and Chemical Engineering
"Giulio Natta", Politecnico di Milano, Milano, Italy
¶Department of Biomedical Engineering
The Chinese University of Hong Kong
Hong Kong SAR, China
‖ypho@cuhk.edu.hk

This chapter is a concise description of the outstanding applications of DNA nanotechnology in the field of Biosensors. Strategies for top-down fabrication of DNA sensors are described and signal transduction mechanisms for sensors are reported.

3.1 Introduction

The general purpose of biosensors is to characterize specific properties of a biological analyte or to characterize an interaction between analytes. Biological affinity to the analyte is typically generated by the molecular probe. This chapter focuses on the use of DNA molecules as probes. Once an interaction between the DNA probe and analyte has occurred, the interaction is transduced into a detectable signal. Depending on the assay formats and signal transduction methods, separation of bound and unbound probes may be required. Typically, heterogeneous assays refer to assays that require separation of bound and unbound species while homogeneous assays do not. Separation can be accomplished through stringency washes to rinse away unbound species provided the probes or analytes are anchored to a suitable support.

Sensitivity, selectivity and sample volume are perhaps the most widely touted measures for the performance of DNA sensors. The definitions of sensitivity and selectivity may vary depending on the context of assays. Generally speaking, sensitivity is defined as the lowest concentration or the minimal amount of analyte that can be detected against the baseline of background. Selectivity refers to the ability of an assay to distinguish between a specific analyte and the background or unspecific analytes. Therefore, selectivity is typically quantified by the "signal-to-noise ratio" (SNR), which is defined as a ratio between the signal produced by the targeted analytes and the noise generated by the interfering species. The sample volume is straightforward and defines the minimal volume of sample required for a particular assay.

Based on the timescales during which measurements are recorded, the assays can also be categorized into equilibrium or real-time assays. In equilibrium assays, the measurements are taken after the steady state is reached. For example, heterogeneous assays require stringency washes. Therefore, the measurement is normally taken when the assay reaches a steady-state equilibrium between probe-analyte binding. For homogeneous assays, provided a proper signal transduction technique, measurements can be recorded real-time as the reaction is ongoing to acquire kinetic information of the reaction.

As an overview to the audience interested in utilizing DNA sensors for various purposes, this chapter aims to present the state-of-art fabrication of DNA sensors, commonly used signal transduction mechanisms, and different sensing strategies for the detection of nucleic acids, proteins, enzymes and their activities as well as biochemical conditions and small molecules.

3.2 Construction of DNA Sensors

To specifically assay the interactions between the DNA probe and the analytes, it is necessary to immobilize DNA onto a detectable component, such as a solid surface or the surface of nanoparticles, forming a DNA sensor without interfering with the base pairing. A variety of methods have been used over the last decade to fabricate and develop miniaturized DNA sensors, given the large breadth of applications found for these devices. The fabrication methods can be generally divided into two major categories: "bottom-up" and "top-down" methods according to the processes involved in creating DNA ordered structures on substrates. A bottom-up approach uses molecular or atomic components as building blocks to build up multi-level structures and self-organizing assemblies based on intermolecular and supramolecular interactions and complex mechanisms [1]. From the bottom-up perspective, the immobilization of DNA is typically achieved through either adsorption or covalent binding, similar to those for the enzyme based biosensors [2, 3]. Adsorption is the simplest as it does not require modifications of the nucleic acids. However, the specificity may be of a concern. On the other hand, covalent conjugation is more specific as there are particular sites that can be modified on the nucleic acids, namely the available bases, the sugar or the phosphate groups to produce derivatives for incorporation of various functional groups such as –SH, –NH2, –COOH, or –biotin to covalently attach with the to-be-conjugated molecules. The conjugation chemistry and protocols for the modification of nucleic acids have been established and reviewed elsewhere [4] and thus are not elaborated herein.

3.2.1 *Top-down fabrication of DNA sensors*

As opposed to the bottom-up approach, top-down approaches correspond to the use of fabrication tools controlled by external experimental parameters to create DNA patterns and functional devices with the desired shapes and characteristics [5]. The ability to ensure a controllable placement of the DNA molecules on desired regions and potentially on large areas makes top-down methods the most widespread approaches for the fabrication of DNA sensors. Among the top-down approaches, various lithography-based techniques have been employed for patterning 2D DNA features.

One established top-down approach is dip-pen nanolithography (DPN), which involves the selective deposition of molecules onto a surface in a regular arrangement of dots or shapes by means of atomic force microscopy (AFM). More specifically, the DNA patterning using the DPN technique involves the use of one or several tips, which are previously dipped in a solution of the molecules to be deposited, commonly referred to as the "ink." The ink is transferred to the substrate when the AFM tip is very close to the substrate (few nanometers). This close contact with the surface enables the AFM tip to release the material of interest according to a predefined pattern obtained with the movement of scanning tips. This method offers the possibility to pattern features with a lateral resolution of around 50 nm, allowing the fabrication of high-resolution DNA nanoarrays [6]. Among the numerous strategies developed for DNA patterning by means of DPN, one of the most employed methods is the direct-write deposition of oligonucleotides on the substrate. This was demonstrated by Demers *et al.* to generate covalently bonded patterns of oligonucleotides on gold and SiO_x substrates [7]. This work also was the first example of functionalization of AFM tips to achieve a successful deposition of DNA. It could be achieved because the surface of silicon nitride tips was modified with a chemical compound, i.e. 3-aminopropyltrimethoxysilane, to promote a good adhesion of DNA ink molecules to the tip surface. By using these modified tips, functionalized oligonucleotides can be patterned on different surfaces with a high-quality resolution, preserving their sequence-specific activity as verified by hybridization with complementary probes. The use of a carrier such as an agarose hydrogel has also been proved to have many advantages over the direct deposition of oligonucleotides in a DPN writing process, resulting in a fast and efficient patterning of oligonucleotide nanostructures, which maintain their biological activity and specificity (Figures 3.1(a)–3.1(c)) [6].

DPN can also be used in the so-called "indirect" approach to obtain a certain pattern, which then directs the immobilization of biomolecules from the solution to the surface. The biomolecule indirect deposition approach requires the ability to anchor these molecules to the substrate through specific interactions, which minimize non-specific surface interactions. For example, the DNA molecules can be immobilized in an ordered arrangement via electrostatic interactions between positively charged amino-terminated arrays printed by DPN and a solution of negatively charged oligonucleotides [8]. Alternatively, immobilization can be achieved via covalent bonds by coupling an amino-functionalized DNA molecules to a DPN-printed pattern with carboxylic acid terminal groups by the formation of an amide bond [9]. Although the DPN technique offers the freedom of printing any kind of patterns and high resolutions with a minimum feature size of 15 nm (for 16-mercaptohexadecanoic acid on gold), the process is serial and therefore is slower than the parallel techniques. To overcome the limitation of printing speed, multiple probes and tip arrays have been developed, thus allowing a large-volume production and high throughput [10, 11].

Another lithographic technique based on scanning probe microscopy is the nanografting method. When compared to other top-down fabrication techniques, the nanografting process shows greater control over the size, geometry and position of biomolecules on the surface. Usually, the nanografting method is employed on surfaces previously modified with self-assembled monolayers (SAMs). The process can be divided in two steps: (a) a pre-existing SAM is removed via nanoshaving using an AFM tip as a razor in the presence of surfactant molecules, having a greater affinity with the substrate than those removed by the tip; (b) once the previous SAM is removed from the addressable area, another different SAM can be obtained over the shaved surface.

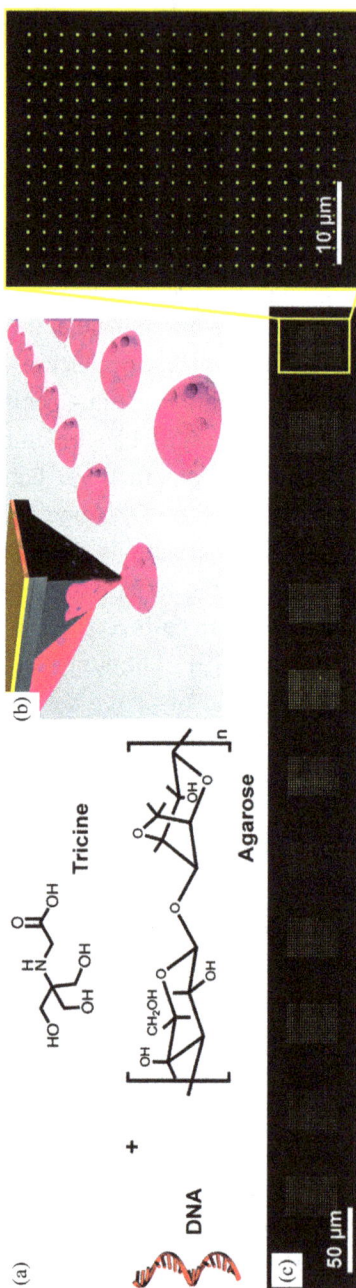

Figure 3.1. (a) Schematic representation of ink and matrix components; (b) an illustration showing the process of agarose-assisted DPN; (c) epifluorescent microscope image of a 15 × 20 array of 500 nm Cy3 labeled oligonucleotide features generated in parallel from a 12-tip cantilever array. Adapted with permission from Ref. [6]. Copyright © (2016) American Chemical Society.

For a successful deposition by nanografting, SAMs have to meet some requirements: (i) the pre-existing SAM must be easily removable by the AFM tip; and (ii) the successive SAM has to be formed very quickly. For these reasons, thiol-gold combination is usually employed for the nanografting, particularly due to the rapid formation of homogeneous monolayers of thiols on gold surfaces. Liu *et al.* demonstrated the deposition of oligonucleotides functionalized with a thiol group by direct nanografting. Due to the presence of pre-existing SAM, it is possible to drive the adsorption of DNA molecules and also to prevent lateral diffusion of DNA, resulting in an excellent feature control [12]. Another advantage of this direct nanografting method is a 3D arrangement of oligonucleotides, which are almost perpendicular to the surface and placed in thin lines wide up to 20 nm surrounded by the pre-existing SAM, thus enabling an easier subsequent hybridization.

Moreover, electron beam lithography (EBL) can also be used for depositing DNA arrays, as shown by Zhang *et al.* [13]. An advantage of this lithographic technique is its compatibility with the normal techniques of microfabrication developed in the electronics field, thus allowing a simple integration of biomolecules in electronic sensors or Bio Nano Electro-Mechanical System (bioNEMS). However, EBL only operates in an atmosphere of ultra-high vacuum, strongly limiting the potential of this technique in manufacturing multi-component nanoarrays.

Another lithographic technique is microcontact printing (μCP), widely used for the deposition of organic molecules and biomolecules on large areas (>cm^2). The μCP process, developed by Kumar and Whitesides [14], usually employs an elastomeric mold. The fabrication of the mold starts from the creation of a negative master mold, on which a liquid prepolymer is poured. Once the polymerization of the prepolymer is obtained through a heating treatment (curing), the mold is separated from the master and ready for use. The mold is then dipped in the solution containing the molecules to be deposited on the surface. After a sufficiently long incubation time which depends on the ink (1–45 min), the mold is dried and pressed on the substrate. Due to the conforming contact between the substrate and the relief parts of the mold, the molecules are transferred to the surface. Once the master is available, the mold fabrication can be completed within a couple of hours. Polydimethylsiloxane (PDMS) is the material of choice for the preparation of the mold, because its low-cost and suitability for normal laboratories. The entire deposition process can also be performed without the need of a clean room. All these factors make this technology very flexible, fast and affordable for most laboratories, contributing to its popularity. Nyamjav and Holz demonstrated the use of μCP as a robust, reliable and inexpensive method to produce predesigned, spatially controlled, high density microarrays of DNA molecules directly on silicon oxide substrates. The coupling of oligonucleotides to a silicon substrate was achieved by silanizing an acrylamide-terminated DNA, resulting in arrayed oligonucleotides with a retained biological function [15].

With the aim of manufacturing low-cost electronic biosensor devices, one of the most promising technologies is inkjet printing. A large versatility is shown by this patterning method, which is also suitable for prototyping. This approach is based on the ejection of a liquid ink through micrometer-sized nozzles under a pressure pulse. It enables the deposition of very small volumes of ink (from pL to nL) in a rapid procedure, achieving high pattern precision, micron-sized resolution and a good reproducibility [16].

3.3 Signal Transduction Mechanisms

The signal transduction for DNA sensors may be achieved through optical, electrical, mechanical or magnetic routes. Optical detection is perhaps the most commonly used method due to the high sensitivity of optical systems and due to the legacy from assays developed with molecular fluorophores and dyes. Optical detection methods are typically based on fluorescence emission, surface plasmon resonance (SPR), scattering, colorimetric change, Förster resonance energy transfer (FRET) or surface enhanced Raman scattering (SERS). This section will review the most widely adopted signal transduction mechanisms based on fluorescence and plasmonics.

3.3.1 *Fluorescence-based sensing*

Fluorophore conjugates, which provide fluorescence or act by changing the spectral properties of the sample, are the basis for a wide range of

DNA sensors. For example, premier Cyanine Dyes (e.g. PicoGreen, SYBR Green) and cyanine dimers or monomers (e.g. TOTO or YOYO family), intercalate DNA to hide their hydrophobic chains from the surrounding aqueous solution. This results in increased fluorescent output by several orders of magnitude [17–19] and has been utilized extensively in genomics research. Aside from the conventional organic fluorophores, the utility of nanoparticles, namely colloidal particles of 5–50 nm in diameter, such as semiconductor nanocrystals and metallic nanoparticles, have fundamentally changed the bioanalytical measurement landscape [20–22]. Luminescent semiconductor nanocrystals, colloquially known as quantum dots (QDs), stand amongst the most exciting research tools in chemistry, physics and biology. These inorganic fluorophore crystals, normally composed of periodic groups of II–IV (e.g. CdSe) or III–V (e.g. InP) material, have demonstrated superior optical properties, such as narrow, symmetric and precise size-tunable emission spectra and broad excitation spectrum. Other commonly discussed benefits of QDs in comparison to organic fluorophores include enhanced fluorescence stability and negligible photobleaching [23]. Consequently, much attention has been focused on the utilization of QD for the construction of DNA sensors [24, 25], particularly for those aiming for multiplexed detection [26, 27].

3.3.1.1 *Förster resonance energy transfer*

FRET is a commonly used strategy in DNA sensors. FRET occurs when the energy of the donor is non-radiatively transferred to a nearby acceptor molecule via a dipole–dipole interaction, leading to a decrease in the donor emission and an increase in the acceptor emission, in such a way that the excited state lifetime of the donor is diminished [28, 29]. The energy transfer efficiency, E, is a function of donor–acceptor separation r, serving as an important measure to the energy transfer process. Equation (1) presents the general definition describing the fraction of excitations transferred from the donor to the acceptor non-radiatively, where k_T is the rate of non-radiative energy transfer and t_D is the excited-state radiative lifetime.

$$E = \frac{k_T}{t_D^{-1} + k_T} = \frac{R_0^6}{R_0^6 + r^6} \tag{1}$$

The Förster radius R_0 refers to the distance at which the energy transfer efficiency is 50%. This is typically used to characterize the FRET relationship of a particular donor–acceptor pair. For most FRET pairs, R_0 is typically a few nanometers, enabling optical measurements of changes in donor–acceptor distance with angstroms resolution, leading to a "nanoruler" probing intramolecular or intermolecular distance.

Ideal FRET occurs when there is appreciable spectral overlap between the emission spectrum of the donor and the absorption spectrum of the acceptor, while crosstalk caused by spectral overlap, of the donor and acceptor emission is minimum. With this request in mind, organic fluorophore FRET pairs often suffer from direct acceptor excitation (excitation of the energy acceptor at the meantime of exciting the energy donor) and crosstalk (spectrum overlap between donor and acceptor emissions), due to their broad adsorption and emission spectra. To this end, there is an emerging effort on involving QDs in the FRET process, since this overcome some of the limitations associated with conventional organic FRET pairs [25, 30–32].

3.3.2 *Plasmonics-based sensing*

Plasmonics is the study of the interaction between an electromagnetic field and free electrons in a noble metal, such as gold or silver [33]. The unique properties of plasmonic materials derive from the plasmon oscillation, i.e. the collective motion of conduction band electrons relative to fixed positive ions. The design of artificial plasmonic structures has shown enormous importance for the production of biosensors, for instance in the form of SPR [34]. SPR can be defined as the phenomenon, which occurs when polarized light, under conditions of total internal reflection, strikes a conducting gold layer at the interface between media of different refractive index such as the glass of a sensor surface (high refractive index) and a buffer (low refractive index). The resonance condition is established if the frequency of incident photons matches the natural frequency of surface electrons of a conducting gold layer oscillating against the restoring force of positive nuclei.

Remarkably, by tuning geometries and order in nanomaterials, it is possible to obtain plasmonic effects showing phenomena such as absorbance and scattering. Such unique features make these materials a hot research topic due to applications such as signal enhancement for bioimaging [35] and biosensing [34]. The commonly used approach to build up nanomaterials with desired plasmonic properties is by conventional top-down photolithography. However, these approaches suffer from high costs and are not suitable to construct 3D structures. In this regard, DNA nanotechnology allows for assembling nanomaterials by bottom-up approaches in order to build up plasmonic metallic nanostructures showing high complexity and hierarchy. In fact, the versatile chemistry of DNA allows fine-tuning of the optical properties of plasmonic nanostructures, such as coupling nanoparticles surface plasmons in chiral assemblies.

For instance, Kuzyk *et al.* [36] designed DNA origami for producing plasmonic structures containing nanoparticles arranged in left- and right-handed nanoscale helices. They employed DNA origami 24-helix bundles having nine helically arranged attachment sites for plasmonic particles coated with single stranded DNA, i.e. gold nanoparticles with a diameter of 10 nm or 16 nm. By bulk measurements, such structures are found to exhibit defined circular dichroism and optical rotatory dispersion effects, which are due to the collective plasmon–plasmon interactions of the nanoparticles positioned.

As an example of an active plasmonic system, Zhou *et al.* demonstrated the possibility to fabricate a plasmonic nanorod that can execute directional, progressive and reverse nanoscale walking on 2D- or 3D DNA origami. The designed walker has an anisotropic gold nanorod and DNA strands as feet that allow it to move on a DNA-origami [37]. More recently, Gur *et al.* fabricated a new class of plasmonic waveguides by thoroughly investigating the effect of several parameters that can influence the attachment yield of oligonucleotide-functionalized AuNPs to six-helix bundle (6-HB) DNA origamis to produce self-assembled plasmonic waveguide precursors. They found out that high ionic strength (300 mM NaCl and 12 mM $MgCl_2$) is necessary to stably hybridize the AuNPs to the tethers on the DNA origami and that hybridization of AuNPs to 6-HBs takes seconds, and aggregate formation is irreversible [38]. Copp *et al.* investigated the effects of dielectric environment and cluster shape on electronic excitations of fluorescent DNA-stabilized silver clusters. Their results suggested the possibility to analyze dielectric environments at length scales of around 1 nm [39].

On the other hand, the plasmonic properties of dimeric nanostructures are also reported. For example, Acuna *et al.* [40] designed a DNA-origami pillar having docking sites to allow for the assembly of nanoparticle dimers apart from each other in a distance of 23 nm. In this manner, self-assembled nanoantenna that allowed for enhanced fluorescence intensity in a plasmonic hotspot of zeptoliter volume was created. They carried out direct visualization of the binding and dissociation of short ATTO 655-labeled DNA strands, as well as the conformational dynamics of a DNA Holliday junction in the hotspot of the nanoantenna, thus showing the compatibility with single-molecule bioassays.

As an alternative to fluorescence detection, some reports deal with Raman detection, a technique which is based on inelastic scattering of monochromatic light, usually from a laser. The laser light interacts with molecular vibrations, phonons or other excitations in the system, resulting in the energy of the laser photons being shifted up or down. Such energy provides useful information about the vibrational modes in the system. Thacker *et al.* [41] carried out plasmonic coupling between two 40 nm gold nanoparticles held with a gap of about 3.3 nm by a DNA origami suitable for Raman detection. They employed such dimers to measure Rhodamine 6G, a model Raman analyte, demonstrating that only in the presence of gold nanoparticle dimers, it was possible to acquire Raman spectra. They also obtained SERS spectra of nanoparticles coated with ssDNA for sequence-specific DNA detection showing an average enhancement of 4–5 orders of magnitude in the scattering cross-section in comparison with bulk measurements — see Figure 3.2(a).

Finally, in a very recent report, Sinha *et al.* [42] showed the possibility to fabricate DNA-mediated gold nanoprism based optical antennas, which can enhance two-photons-imaging capability in the 1100 nm biological II window. In the process of two-photon fluorescence (TPF), nanoparticles or organic molecules absorb two low energy photons and emit a single excited photon at an energy level higher than that of the absorbed photons. The report demonstrated enhanced second harmonic generation (SHG) and TPF properties by several orders of magnitude via DNA-driven plasmon-coupled organization into gold nanoprism assembly structures using human Hep G2 liver cancer cells as a model.

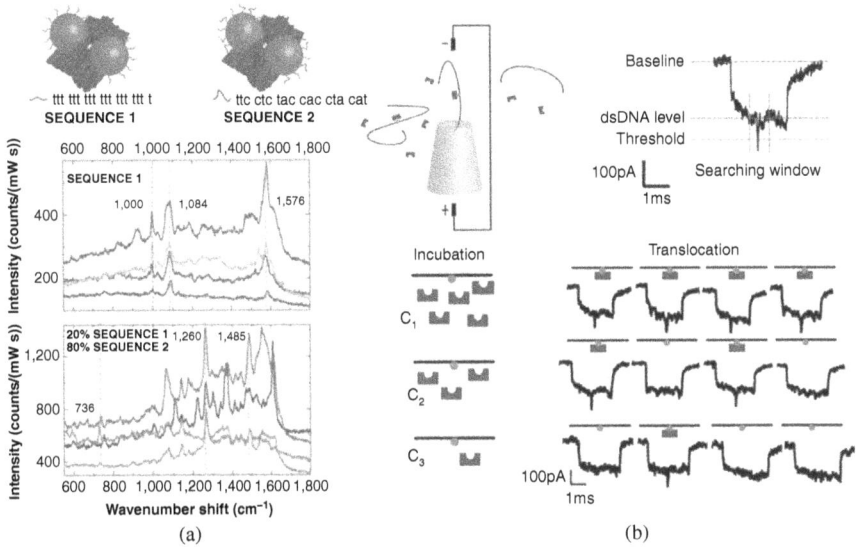

Figure 3.2. (a) DNA origami based surface-enhanced Raman scattering for DNA sensing. Gold nanoparticles — plated on a gold coated silicon surface — are coated with two different types of DNA sequences (sequence 1 is made of 19 thymines and sequence 2 adenine and cytosine in an 80:20 ratio). SERS spectra can distinguish the two different analytes, since SERS spectra of sequence 1 gives peaks corresponding to the DNA backbone (1000 cm^{-1}) and thymine (1084 cm^{-1}), whereas SERS spectra of sequence 2 shows peaks corresponding to adenine (736, 1260 and 1485 cm^{-1}) and cytosine (1260 and 1485 cm^{-1}). Reprinted with permission from Macmillan Publishers Ltd: Nature Communications (Ref. [41]), copyright © (2014). (b) At the left, representation of DNA carrier (black line) and protein complexes (black line with cuboid attached) translocating through a nanopore driven by the electric field. At the right, carrier translocation event showing an extra second level current drop in the middle of the event, due to the targeted protein binding. By decreasing protein/carrier concentration ratio ($c_3 < c_2 < c_1$), less occupied events are found. Reproduced from Ref. [47]. Copyright © (2016) American Chemical Society under Creative Commons Attribution (CC-BY) License.

3.3.3 *DNA nanopore-based sensing*

DNA nanopore-based sensing represents one of the most interesting applications of DNA-based nanostructures. Under an application of constant voltage, nanopore-based detection measures the ionic current between two electrolyte solutions separating by a single nanoscale pore. As individual molecules translocate the pore, the flow of ions passing through the pore reduces and therefore the measured current [43]. The first example of nanopore-based sensing was presented by Bayley and coworkers with the α-hemolysin protein [44], a nanopore from bacteria that causes lysis of red blood cells. The report showed the capability of nanopore-based sensing for DNA sequencing, since all four bases can be identified by the measured ionic current. Furthermore, research efforts have permitted the understanding of how DNA molecules interact with nanopore system during the translocation process. When bound with analytes through covalent/non-covalent interactions, DNA hybrids produce characteristic current signals when translocated through nanopore which permit analytes identification. Researchers have also employed DNA carriers molecules, long DNA strands which have specific binding sites for target analytes in solution. Such DNA constructs are able to interact with analyte (typically a protein) and, upon translocation through the nanopore, specific current signals of the DNA–analyte complex are measured. Analyte sensing can be carried out via signature current events generated by different DNA–analyte interactions, such as DNA hybridization, DNA aptamer–analyte binding, DNA metal ion binding, DNA protein interactions and host–guest interactions [43].

In addition to α-hemolysin protein, synthetic DNA-origami nanostructures have also been applied as a nanopore by two different approaches: (1) by trapping a DNA-origami structure at the mouth of a solid state nanopore, and (2) by an insertion of a DNA-origami structure, coated with hydrophobic moieties, into a lipid bilayer.

As an example of the first approach (i.e. inserting DNA nanostructures inside solid state nanopore), Bell et al. reported a strategy to repeatedly insert and eject funnel shaped DNA origami designed with a constriction of 3 × 3 double-helix widths (7.5 nm × 7.5 nm) through solid-state nanopores with diameters around 15 nm for detecting model DNA analytes, such as λ-DNA molecules [45]. Hernández-Ainsa et al. combined DNA-origami structures with glass nanocapillaries to reversibly form hybrid DNA-origami nanopores [46]. As a very recent example, Kong et al. [47] combined solid-state nanopores and DNA carriers to quantify nanomolar level of protein concentration. They could determine protein concentration by measuring the fraction of DNA carriers which showed a secondary signal indicating the presence of a protein. They used two model systems, namely biotin–streptavidin and digoxigenin–antidigoxigenin — see Figure 3.2(b). The working protein concentrations were at nanomolar range, and microfluidic chips with ~10 μL volume were used, so the total amount of protein needed was only 10 fmol.

The second approach is conducted by the insertion of a structure into a lipid bilayer. In this scenario, it is possible to mimic membrane proteins which insert into cell membranes for controlling the transport of molecules and ions. The charged phosphate backbone of DNA is hydrophilic and so the DNA nanostructure must be decorated with hydrophobic chemical groups to overcome the large free energy barrier to form a hole in the lipid bilayer. In a fundamental report, Langecker *et al.* linked cholesterol tagged oligonucleotides to a stem and barrel DNA-origami structure which formed a pore 2 nm wide and 42 nm in length. Such nanostructured pores were capable of insertion inside lipid vesicles with the desired orientation [48].

3.4 Detection of Nucleic Acids

DNA sensors for detection of nucleic acids usually rely on the specific Watson–Crick base-pairing between polynucleotides. Some of the first were simply labeled oligonucleotides which could be hybridized to a target sequence. Although these were robust they had low resolution and were incapable of detecting minor stretches or variations in nucleic acids. Specific detection of small sequence stretches or even single nucleotide variations required highly specific DNA sensors in combination with powerful amplification strategies. One of the first examples of such highly specific DNA sensors was the so-called Padlock probes [49]. Padlock probes are DNA oligonucleotides consisting of two target-complementary DNA sequences, linked by a freely designable DNA sequence typically used for probe identification. Upon hybridization to the target sequence the 5′- and 3′-ends of the padlock probe are brought into proximity allowing the open circular DNA structure to be converted to a closed circular DNA structure by the action of a DNA ligase [49]. Due to the helical structure of DNA, padlock probes are topologically linked to the target DNA after hybridization and ligation. Padlock probes, therefore, allow stringent washing conditions and thus reduced background staining despite a short hybridization sequence [50, 51]. This, combined with the specificity provided by the ligation event, which is sensitive enough to distinguish single base differences [52] enabled detection of single nucleotide variations in the alpha satellite of centromeres in human chromosomes [53].

The significant advantage of padlock probes is due to their capability to change from an open circular DNA structure to a closed circular DNA structure. This closed DNA structure can be used as template for subsequent amplification strategies such as Rolling Circle Amplification (RCA) [54–58], or Polymerase Chain Reaction (PCR) [59, 60]. To better allow multiplexing, the strategy of padlock probes was further modified in what was named Molecular Inversion Probes (MIP) [59] and Connector Inversion Probes (CIP) [61]. Both types of probes closely resemble padlock probes but are designed to hybridize with either a single nucleotide gap (MIP) or a larger gap (CIP) between the 5'- and 3'-end. Upon hybridization, the gap is filled by a DNA polymerase upon addition of either the specific nucleotide complementary to the single nucleotide gap (for the MIP) or all four nucleotides (for the CIP). The readout was performed using genotyping arrays with sequences complementary to the linker sequence or using sequencing [59, 61, 62].

Detection of RNA targets using padlock probes in combination with RCA [55, 63, 64] or hyperbranched RCA (hRCA) [65–67] have also been reported. However, although DNA ligases can ligate DNA using an RNA template it is less efficient than using a DNA based template [68–72]. Therefore, different solutions have been proposed to circumvent the use of RNA as a template for the ligation. One solution was to use preligated closed circles primed from the RNA-target [67, 73, 74]. A second method was to use a DNA sensor termed turtle probe designed to form a secondary hairpin structure bringing the 5'- and 3'-end into proximity facilitating ligation to form a closed circle which could serve as template during RCA primed from the target RNA [75, 76]. The last technique was to reverse transcribe the RNA to cDNA using an LNA primer, degrade the RNA to make a single stranded target, and detect the cDNA with padlock probes [77].

A different group of DNA sensors for detection of RNA are based on DNAzymes, also called deoxyribozymes, which are DNA sequences with catalytic functions. Unlike RNAzymes, also referred to as hammerhead ribozymes, which have been found widely distributed in the nature [78], ranging from viroids and viral satellites [79, 80] to humans [81], no naturally occurring DNAzymes have been reported. DNAzymes are a result of *in vitro* selection and the first reported catalyzed Pb^{2+}-dependent cleavage of RNA [82]. Soon after several others were reported [83, 84], including the general purpose RNA cleaving 10–23 and 8–17 DNAzymes [85] and the horseradish peroxidase (HRP) mimicking DNAzyme [86, 87]. DNAzymes have formed the basis for many different sensors using different read-outs e.g. electrochemical [88, 89], fluorescent [90], or colorimetric [91].

3.5 Detection of Proteins

The development of synthetic DNA/RNA aptamers via the Systematic Evolution of Ligands by Exponential Enrichment (SELEX) technology [92–94] have led to the development of aptamers capable of recognizing specific proteins [94]. The development of aptamers have illustrated that protein detection can be performed using DNA sensors in assays resembling antibody sensors without the need for antibody development. Moreover, as described for nucleic acid detecting sensors, the use of DNA sensors for protein detection allow the use of a variety of readouts e.g. direct labeling of the aptamer with fluorescence or biotin followed by direct or indirect visualization using streptavidin-conjugated Phycoerythrin (PE) or QD [95]. Alternatively, different amplification strategies have been used such as RCA primed directly from the aptamer [96–101] or PCR of the aptamer [102, 103].

DNA sensors in the form of aptamers have been used for directly sensing biomolecules in solution relying on proximity ligation [104] or proximity extension [99]. Proximity ligation or extension is based on the principle that two or more target binding sensors e.g. aptamers have to be brought in proximity before an amplification (e.g. RCA or PCR) leading to detectable signal can take place [104, 105]. Although DNA sensors in the form of aptamers seem appealing for proximity assays, the lack of suitable aptamers has led to the use of DNA labelled antibodies rather than the DNA aptamers [106, 107].

Another promising family of sensors for detection of proteins in solution is based on structure-switching aptamers. Several different types of structure-switching aptamer based DNA sensors have been developed using various strategies. One type changes into an exonuclease resistant structure upon binding of the target protein thus preventing degradation of the DNA aptamer and resulting in a fluorescent signal by addition of an intercalating fluorescent dye [108]. Although beautiful in its simplicity, this approach may be challenged by its specificity in the presence of DNA binding proteins. A second nuclease based structure-switching aptamer system changes from a hairpin shaped structure into a partly single-stranded stretch of DNA upon target protein (thrombin) binding to an aptamer. The single-stranded stretch of DNA can hybridize to a fluoro-phore–quencher conjugated probe generating a double-stranded substrate for a nicking endonuclease that can cleave the probe and thus causing an increase in fluorescence through separation of the quencher and fluoro-phore [109]. A different approach was demonstrated by Yang *et al.* who developed a structure-switching aptamer that could be circularized only upon specific binding of its protein target generating the template for a subsequent RCA [110]. Lastly, a universal structure-switching aptamer based system has been presented in which an aptamer and a padlock probe compete for binding to a specific oligonucleotide. Upon binding of the target protein to the aptamer, the oligonucleotide is free and can serve as a template for the padlock probe that can be amplified using RCA [111]. This system presents the significant advantage of being easily adaptable to the detection of other biomolecules although the target binding sequence has to be changed for each new target detected.

3.6 Detection of Enzymes

DNA sensors for detection of enzyme activity may find use in many different fields of basic or applied science for e.g. basic enzyme characterizations, drug screening, prognostics or diagnostics [112–114]. One of the most straightforward advantages of DNA sensors tailored for measuring enzyme activity, is that they can easily be converted to "real time sensors" allowing investigation of kinetic parameters and the acquisition of large data sets from single reactions.

Examples of DNA based real time sensors for monitoring enzyme activities are a large variety of DNA oligonucleotides coupled to quencher-fluorophore or other FRET pairs designed in such a way that the two parts are either brought together or separated upon reaction with the target enzyme. In this manner interaction with the target enzyme results in either gain or loss of signal depending on the specific design of the sensor. Examples of the reporter fluorophores include both organic fluorophores and semiconductor nanocrystals (e.g. QDs). Moreover, readout systems based on carbon nanoparticles, or gold nanorods have also been reported. The different sensor systems have been demonstrated for detection of different enzyme target activities including endonuclease induced DNA cleavage [115–120] or binding (the latter by using single molecule FRET [121]), DNA polymerases [122], DNA repair activities [123–126] and various topoisomerases [127–130]. These types of sensors excel by the ease and speed by which large data sets can be obtained, and for this reason they have been demonstrated of particular use for kinetic analyses. However, with regard to sensitivity they are inferior to more tedious protocols involving several amplification steps as e.g. RCA-based protocols (see below).

Other examples of real time DNA sensors for monitoring enzyme activities are the so-called label free sensors. In one setup, the detection in the DNA sensor was transduced into electronic signals by a monolayer graphene field-effect sensor containing DNA oligonucleotides coupled to monolayer graphene strips that allow real-time measurement of human topoisomerase I-mediated cleavage and ligation [131].

Already at this early stage of their development, real time sensors have demonstrated their superiority when it comes to basic investigation of enzymatic reactions including the effect of various inhibitors or potential small molecule drugs, and as such real-time sensors hold promise for future drug screening programs [132]. Furthermore, the inclusion of real time sensors in existing assays, such as ELISA have enabled simultaneously detection of protein amount and enzymatic activity in extract from cells and tissue [133]. Several examples of using real time DNA sensors for measuring various enzyme activities in crude extract from biological samples e.g. tissue or cells have also been reported [124, 126, 128, 130]. Such measurements may be of importance for diagnostic or prognostic purposes of specific enzyme activities that may be used as biomarker for disease and/or treatment response.

In cases where the available samples are limited or the enzymatic marker of interest is rare, the optical or electrochemical signals generated directly by the enzymatic reaction with a real time DNA sensor may be insufficient for detection. Hence, to obtain maximal detection levels the sensor module may be combined with one or more signal amplification steps. Sensors composed of DNA provides the benefit of being compatible with a large number of different amplification methodologies including PCR [134], RCA [112], Loop Mediated Amplification [135], Strand Displacement Amplification [136] and many more [137]. Of these, RCA has been used for signal enhancement in relation to the analyses of various enzyme activities. The method presents the advantage of being an isothermal reaction keeping an one-to-one stoichiometry between the RCA product and the original DNA molecule that is amplified. Moreover, RCA generates a long tandem repeat product that can be detected in bulk sample or at the single molecule level. Single molecule visualization techniques include AFM [138–141], Transmission Electron Microscopy (TEM) [142–144], and fluorescence microscopy (following labelling of the RCA product with fluorescence probes or nucleotides) [145–148]. Methods for bulk sample analyses include colorimetric readout by employing enzyme or nucleic acid catalyzed color formation e.g. oxidation of TMB (3,3′,5,5′-tetramethylbenzidine), a chromogen that generates a blue color when oxidized. As mentioned above, because RCA is performed using an isothermal protocol, the reaction follows simple linear kinetics. This is in contrast to the exponential amplification of PCR and enables straightforward quantitative measurements of enzyme activities without reference to multiple standard samples [54, 145, 148, 149]. In order to allow enzyme activity detection to be combined with signal amplification by RCA we recently developed a family of oligonucleotide based substrates that are circularized only upon reaction with their specific target enzymes [145, 146, 148, 150]. By converting the DNA circle product of each enzymatic reaction to a long tandem repeat RCA product this system allowed direct analysis of endogenous enzyme activity at the single catalytic event level. Taking advantage of a sensor oligonucleotide that reacts only with one specific partner enzyme i.e. a specific topoisomerase or recombinase [145, 146, 148] to generate a DNA circle that can template RCA, the Rolling circle Enhanced Enzyme Activity Detection (REEAD) setup is suitable for detecting enzyme activity in crude samples. Moreover, REEAD has proven robust, allowing specific and multiplexed detection of different enzymes in extracts from small samples of human cells or, when combined with droplet microfluidics, even in single cells (Figure 3.3) [151]. Since REEAD is not a real-time method it is less well suited for kinetic analyses than the above described real-time sensors. However, due to the multiple signal amplification steps incorporated in the REEAD the method is typically superior to the real time sensors with respect to sensitivity showing a detection limit of 0.1 fmol of purified enzyme [145]. In case high sensitivity is not needed a simplified but high throughput version of REEAD using real time measurement of the RCA have been demonstrated [152].

Besides basic analyses of the target enzymes such as topoisomerases or recombinases, the REEAD assay may be useful for analyzing the molecular background behind development of disease or drug resistance in e.g. cancer cell populations. Some of the enzymes that have been monitored by REEAD are members of the DNA topoisomerase family of which several are cellular targets of routinely used anti-cancer chemotherapeutics [153–155]. Many of these drugs act by converting the topoisomerase activity to a cell poison and, consequently, their anti-cancer effect depends on the topoisomerase activity level in the cells. For this reason, the topoisomerase specific REEAD assay was envisioned as a powerful future cancer-prognostic tool suitable for analysis of small biopsy samples [145]. More recently, the development of a REEAD setup specific for detecting the activity of topoisomerase I expressed in the malaria causing *Plasmodium* parasites [148, 156] opened up for the application of REEAD in diagnosis of infectious diseases in human or life stock. Combined with droplet microfluidics the multiple signal

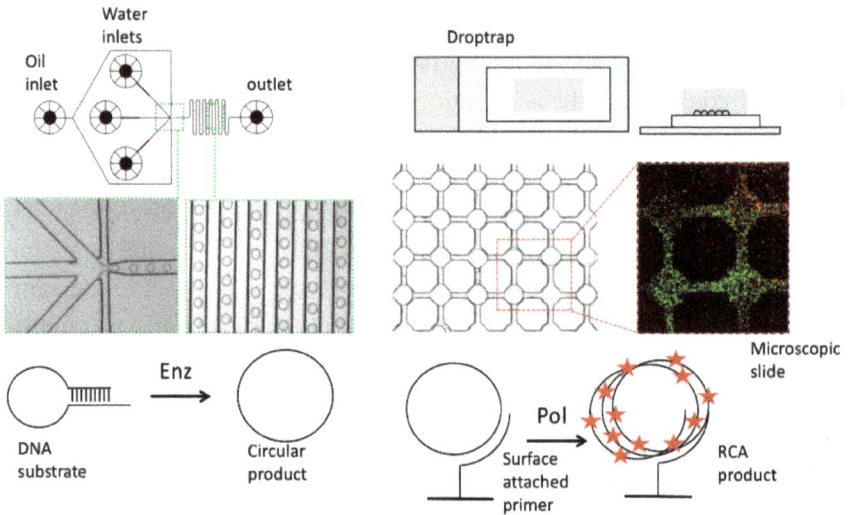

Figure 3.3. Diagnosis by RCA based sensor systems. Upper panel (left) shows the extraction unit composed of a droplet microfluidics chip. An oil phase is injected in an oil inlet while water phases containing the sample to be analyzed, DNA substrate and lyses buffer are injected in the water phase inlets. pL water in oil droplets are generated when the water phase is broken up by the oil phase. These droplets function as micro-reactors and ensure enhanced mixing and reaction kinetics when led though a serpentine channel before they are collected from the outlet. Lower panel (left) shows the circularization of a given DNA substrate that occurs in the droplets. Thereafter the droplets are dispersed and their content exposed to a primer functionalized glass slide using a drop-trap device shown at the upper panel (right). After RCA reaction products can be visualized in a fluorescence microscope (middle panel, right). The lower panel shows the reaction going on the primer functionalized glass surface, where the generated circles bind the primer. Thereafter RCA is performed by an added polymerase giving rise to a long tandem repeat product that can be visualized at the single molecule level.

enhancements imposed by the REEAD setup allowed for an impressive detection limit of 0.06 parasites/μL in a single drop of unprocessed blood. This detection limit was superior to the best PCR protocols and allowed detection of malaria even in saliva from infected individuals [148]. Since the development of the *Plasmodium* topoisomerase I specific REEAD setup, increasing amount of REEAD assays with specificity towards pathogen-, virus- or bacteria-specific DNA interacting enzymes have been established [150].

3.7 Detection of Biochemical Conditions and Small Molecules

Temperature, pH, ionic strength or various co-factors are parameters that affect most chemical reactions and biological processes. Accurate measurement of these parameters is therefore of major importance in medicine and in most branches of science and technology. DNA has turned out to be a prominent material for the generation of sensors that can measure such environmental factors. One reason for this is that the DNA molecule does not have a static structure.

In normal double-stranded DNA, the two strands are kept together by hydrogen bonds between nucleotide bases. Depending on the base composition and environmental factors the DNA duplex can attain different conformations including the common right handed B-DNA conformation or the left handed Z-DNA conformations. Also, the bases of DNA can stack in alternative conformations to generate triplex or quadruplex structures. Breakage of the hydrogen bonds to separate the DNA strands or alterations in DNA conformations can be facilitated by changes in the surrounding aqueous solution, most notably by changes in the temperature, pH or the ionic strength of the aqueous solution. By the introduction of various modifications e.g. fluorophore–quencher pairs it has been possible to demonstrate the functionality of a variety of DNA sensors for monitoring different chemical or physical characteristics of the surroundings.

For the generation of thermometers it has been exploited that each duplex DNA structure in a given solution is characterized by a specific melting temperature (T_m), which is the temperature where half of the molecules have melted and generated two DNA strands. The fraction of DNA molecules melted at a given temperature reflects an equilibrium process between DNA melting and re-annealing (hybridization) of the two strands. T_m for a given DNA molecule depends on the number of hydrogen bonds and therefore on the number and composition of base-pairs. A temperature sensor is thus inherited in the hybridization kinetics of DNA molecules. Consistently, nanoscale "DNA-thermometers" have been presented taking advantage of the hybridization kinetics of DNA molecules [157–160]. So far the DNA-thermometers have been based on the so-called "molecular

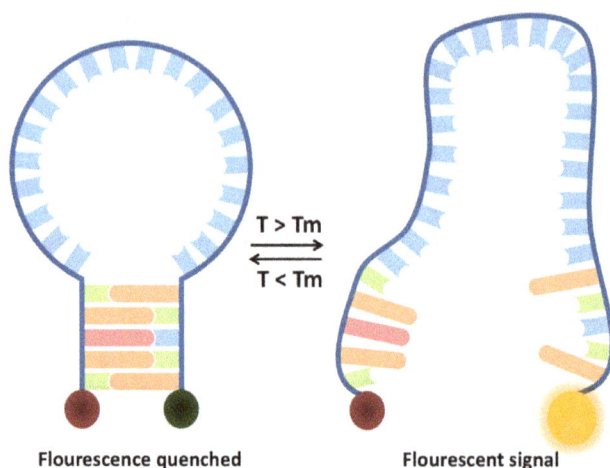

Figure 3.4. Schematic illustration of a DNA hairpin end-labeled with a donor–acceptor pair of dyes. In the hairpin conformation, the acceptor (red) quenches the light emitted from the donor (green). When the temperature increases above the melting temperature of the hairpin (T_m) the hairpin melts, and separation of the dyes results in a fluorescent signal.

beacons" [158] consisting of DNA molecules able to form DNA hairpin structures. Hairpins are generated from single stranded DNA which has complementary sequences at the ends, allowing formation of a DNA hairpin as a consequence of intra-strand hybridization between complementary sequences (Figure 3.4). In a DNA hairpin the duplex region generated upon hybridization forms the stem of the hairpin and the nucleotides between make up the hairpin loop. At a specific temperature an equilibrium will form between closed hairpins and single-stranded DNA molecules. At temperatures below T_m of the specific DNA molecule, most molecules will form a hairpin, but as the temperature rises above T_m the hairpins will melt, and more molecules will be single-stranded [159]. Both the stem length and sequence of the stem is important for the T_m of the hairpin. Moreover, the composition of the surroundings e.g. ionic strengths and pH will affect the T_m and, hence, these factors need to be taken into consideration when measuring temperature using DNA-based sensors.

In the molecular beacon the hybridization kinetics is followed by FRET between a fluorophore and a quencher located at the 5′- and 3′-end of the DNA hairpin, respectively [158]. When $T < T_m$, hairpin formation will ensure that the fluorophore is located close to the quencher in the hairpin stem causing low fluorescence, but as the temperature increases DNA molecules become single-stranded and the quencher is separated from the fluorophore causing higher levels of fluorescence. The fluorescent signal will increase with increasing temperatures giving a sigmoid melting curve [158–160].

The applicability of molecular beacons functionalized with terminal fluorophore–quencher pairs as temperature sensors has been demonstrated using a hairpin with a five base-pair stem consisting of four G–C base-pairs with a centrally located A–T base-pair and a loop of twenty T's [159]. Using this sensor it was possible to follow temperature changes in steps of 1° in the range from 37°C to 42°C and from 42°C to 32°C in many repeated cycles by measuring fluorescence emission resulting from melting of the beacon structure. In the same study it was demonstrated that the temperature range in which a given molecular beacon has its optimal effect as a thermometer can be changed simply by changing the base sequence of the hairpin. Note here, that the temperature range where a given hairpin has its optimal use as a thermometer is the temperature range where the melting curve is steepest. The steeper the melting curve is, the more accurate the thermometer will be, but at the same time the range, where the thermometer can be used will be smaller. As discussed by the authors, this can be compensated for by the simultaneous use of structurally different hairpins, which vary in their optimal temperature range and are coupled to different fluorophore–quencher pairs.

A separate work has demonstrated the application of molecular beacons as intracellular thermometers [160]. In this study the artificial DNA-like molecule termed L-DNA was used for the construction of the molecular beacon that sensed the temperature. The advantage of L-DNA for intracellular measurements of temperature is that L-DNA can neither hybridize with natural DNA nor bind proteins, and it is resistant to intracellular enzymatic degradation. These are all characteristics, which are essential when applying the thermometer to living cells [160]. The L-DNA thermometer was introduced into different cells that were kept at various temperatures, and measurements of fluorescence intensity by confocal microscopy demonstrated similar response curves as obtained with buffer.

Thermal therapy where cells are killed due to a high local temperature increase is a promising method for cancer treatment [161]. However, this method has been hampered by the lack of reliable means to measure the temperature in the organism at the single cell level. The L-DNA thermometers have been used to monitor temperature inside cells that have been added Pd nanosheets [162] and irradiated with a 808 nM laser to increase the intracellular temperature. In these experiments the L-DNA thermometers were able to give reliable measurements of the intracellular temperature. These thermometers may therefore be valuable tools in future treatment of cancer using thermal therapy.

As mentioned above, besides temperature, hybridization of nucleotide bases depends on the ion strength and pH of the surrounding aqueous solution [159]. A higher ionic strength and a higher pH increase the stability of the DNA double-helix due to a partly neutralization of the repulsive electrostatic interactions between the phosphates in the DNA strands. In the work with the L-DNA thermometers, it was demonstrated that the melting curves of the used hairpins did not change very much under the normal intracellular variations in ionic strength and pH, which have been suggested to vary between 150 and 200 nM of monovalent salts and pH 6.8 to 7.4 [160, 163, 164]. However, in the work by Johnstrup *et al.* a severe effect was observed when the ionic strength was increased from 0 to 1000 mM of NaCl. As suggested by the authors the ionic strength can be used to fine tune DNA thermometers to obtain a desired temperature range or alternatively, DNA hairpins can be used to directly measure the ionic strength in solutions, where the temperature is kept constant. In the same way DNA hairpins can be used as pH-meter in situations, where temperature is kept constant [159].

Other examples of sensors for monitoring pH in living cells have been presented by the Krishnan group who described the utilization of a mismatched duplex with a functional C-rich domain that forms an intramolecular i-motif at acidic pH. In this setup donor and acceptor fluorophores inserted in the C-rich domain were used as reporters for pH dependent conformational changes [165]. By changing the sequence of the pH sensitive domain of this type of sensors it was possible to tune the pH sensitive regimens and a family of sensors spanning ranges from pH 4 to 7.5 was created [166]. The sensors were used to probe pH of early endosomes and the trans-Golgi network in the same cell and in real time [167]. Moreover, by fusion to recombinant antibodies it was possible to deliver the pH sensors along the endocytic pathways to specific organelles [168, 169]. In another design Amodio *et al.* [170] utilized triplex-based DNA strand displacement strategies that can be triggered and finely regulated at either basic or acidic pHs to monitor pH in the surrounding solution, while aggregation of poly-dA functionalized gold-nanoparticles by pH dependent A-motif formation allowed a colorimetric readout of pH [171].

On top of the above mentioned pH sensors numerous nanoswitches or — sensors have been designed to measure the chemical composition of their surroundings. These include a mechano-electronic DNA switch designed by the Sen group [172]. This switch binds Hg^{2+} in T–T mismatch regions in a manner that leads to a conformation change, which can be measured by FRET. In one of the earlier examples Seeman and coworkers used the transition between the B and Z form of DNA in a rigid double-crossover molecule to monitor the chemical composition in the surrounding buffer [173].

The identification of increasing numbers of small molecule biomarkers with relevance for treatment or diagnosis of human diseases [174–177] have led to the development of an increasing number of assays for detection of such analytes. Oligonucleotides with unique target-recognition elements, namely the so-called aptamers, have been introduced [92–94, 178]. Among other things, aptamers have been used as key-lock systems for nanocarriers and they were demonstrated to facilitate release of a reporter molecule from a DNA icosahedron upon binding of a small molecule chemical trigger (cdGMP) [179].

Aptamers can also be combined with the above mentioned RCA based signal enhancement methods. Indeed, a promising family of assays for detection of biomolecules in solution takes advantage of clever designed structure-switching aptamers that can be circularized and thereby template RCA only upon target binding. One example on this was the RCA based detection of the small molecule ATP by Cho and coworkers [96].

Table 3.1. DNA sensors for the detection of various kinds of biomolecules and biochemical conditions.

Assays	DNA Probes	Signal Transduction Mechanisms	Timescale	References
DNA/RNA				
DNA	ssDNA	Optical	Equilibrium	[50–62]
RNA (cDNA)	ssDNA	Optical and electrochemical	Equilibrium	[55, 63–67, 73–76, 88–91]
Proteins				
Proteins	ssDNA (aptamer) and ssDNA conjugated antibodies	Optical and electrochemical	Equilibrium	[98, 99, 101, 104–111]
Enzyme Activities				
Endonuclease detection	ssDNA dsDNA	Optical	Real time	[115–120]
DNA polymerases	dsDNA	Optical	Real time	[122]
DNA repair activities	ssDNA dsDNA	Optical	Real time	[123–126]
DNA topoisomerases	ssDNA dsDNA	Optical and electronic	Real time	[127–130]
DNA topoisomerases and recombinases	ssDNA	Optical and mechanical	Equilibrium	[145, 146, 148]
Biochemical Conditions				
Temperature	ssDNA	Optical	Real time	[159, 160]
pH	ssDNA ssDNA	Optical	Real time	[165–167, 170]
Ions	dsDNA	Optical and electronic	Real time	[172, 173]
Small molecules	ssDNA	Optical		[96, 179]

3.8 Conclusion and Future Perspective

The development of DNA nanosensors is currently a vibrant area which has shown significant promise on detecting an array of biomolecules and biochemical conditions, as outlined in Table 3.1. Aside from the excitement of utilizing DNA sensors in studying fundamental reaction kinetics or single cell characteristics, there exist a multitude of opportunities for promoting DNA nanosensors into clinical diagnostics. However, it is imperative to consider the challenges of maintaining the functionality in biological crude samples, and the ease of detection. Most DNA nanosensors have been validated in the absence of body fluidics where there may be non-specific biological components interfering to the sensor specificity and sensitivity; The emerging detection approaches through mechanisms of, for example SPR, scattering, or surface enhanced Raman scattering, are highly sensitive and specific, however, highly dedicated equipment is often required and is thus not suitable for the clinical setting. As the field evolves, there will be undoubtedly a great number of strategies developed to conquer the above-mentioned challenges, and we may see DNA nanosensors entering the clinics in the foreseeable near future.

Acknowledgements

The authors would like to acknowledge the support from the Start Up Fund and the Direct Grant provided by the Chinese University of Hong Kong. We would also like to thank Dr. Sissel Juul Jensen in granting the use of her results in Figure 3.3.

References

1. Sakakibara, K., Hill, J.P., and Ariga, K. (2011). Thin-film-based nanoarchitectures for soft matter: Controlled assembly into two-dimensional worlds. *Small*. **7**(10), 1288–1308.
2. Scouten, W.H., Luong, J.H.T., and Brown, R.S. (1995). Enzyme or protein immobilization techniques for applications in biosensor design. *Trends in Biotechnol*. **13**(5), 178–185.
3. Sassolas, A., Leca-Bouvier, B.D., and Blum, L.J. (2008). DNA biosensors and microarrays. *Chem. Rev.* **108**(1), 109–139.

4. Hermanson, G.T. (2013). Nucleic acid and oligonucleotide modification and conjugation. In *Bioconjugate Techniques* (3rd edition). Academic Press: Boston, 959–987.

5. Gates, B.D., *et al.* (2005). New approaches to nanofabrication: molding, printing, and other techniques. *Chem. Rev.* **105**(4), 1171–1196.

6. Senesi, A.J., *et al.* (2009). Agarose-assisted dip-pen nanolithography of oligonucleotides and proteins. *ACS Nano.* **3**(8), 2394–2402.

7. Demers, L.M., *et al.* (2002). Direct patterning of modified oligonucleotides on metals and insulators by dip-pen nanolithography. *Science.* **296**(5574), 1836–1838.

8. Nyamjav, D., and Ivanisevic, A. (2005). Templates for DNA-templated Fe3O4 nanoparticles. *Biomaterials.* **26**(15), 2749–2757.

9. Demers, L.M., *et al.* (2001). Orthogonal assembly of nanoparticle building blocks on dip-pen nanolithographically generated templates of DNA. *Angew. Chem. Int. Ed. Engl.* **40**(16), 3071–3073.

10. Bullen, D., *et al.* (2004). Parallel dip-pen nanolithography with arrays of individually addressable cantilevers. *Appl. Phys. Lett.* **84**(5), 789–791.

11. Brown, K.A., *et al.* (2013). A cantilever-free approach to dot-matrix nanoprinting. *Proc. Natl. Acad. Sci. USA.* **110**(32), 12921–12924.

12. Liu, M.Z., *et al.* (2002). Production of nanostructures of DNA on surfaces. *Nano Lett.* **2**(8), 863–867.

13. Zhang, G.J., *et al.* (2004). Patterning of DNA nanostructures on silicon surface by electron beam lithography of self-assembled monolayer. *Chem. Commun.* (7), 786–787.

14. Kumar, A., and Whitesides, G.M. (1993). Features of gold having micrometer to centimeter dimensions can be formed through a combination of stamping with an elastomeric stamp and an alkanethiol "ink" followed by chemical etching. *Appl. Phys. Lett.* **63**(14), 2002–2004.

15. Nyamjav, D., and Holz, R.C. (2010). Direct patterning of silanized-biomolecules on semiconductor surfaces. *Langmuir* **26**(23), 18300–18302.

16. Arrabito, G., and Pignataro, B. (2012). Solution processed micro- and nano-bioarrays for multiplexed biosensing. *Anal. Chem.* **84**(13), 5450–5462.

17. Furstenberg, A., *et al.* (2006). Ultrafast excited-state dynamics of DNA fluorescent intercalators: New insight into the fluorescence enhancement mechanism. *J. Am. Chem. Soc.* **128**, 7661–7669.

18. Rye, H.S., and Glazer, A.N. (1995). Interaction of dimeric intercalating dyes with single-stranded DNA. *Nucleic Acids Res.* **23**(7), 1215–1222.

19. Rye, H.S., *et al.* (1992). Stable fluorescent complexes of double-stranded DNA with bis-intercalating asymmetric cyanine dyes — properties and applications. *Nucleic Acids Res.* **20**(11), 2803–2812.

20. Niemeyer, C.M. (2001). Nanoparticles, proteins, and nucleic acids: Biotechnology meets materials science. *Angew. Chem. Int. Ed. Engl.* **40**(22), 4128–4158.
21. Penn, S.G., He, L., and Natan, M.J. (2003). Nanoparticles for bioanalysis. *Curr. Opin. Chem. Biol.* **7**(5), 609–615.
22. Katz, E., and Willner, I. (2004). Integrated nanoparticle-biomolecule hybrid systems: Synthesis, properties, and applications. *Angew. Chem. Int. Ed. Engl.* **43**(45), 6042–6108.
23. Wu, X.Y., *et al.* (2003). Immunofluorescent labeling of cancer marker Her2 and other cellular targets with semiconductor quantum dots. *Nat. Biotechnol.* **21**(1), 41–46.
24. Gerion, D., *et al.* (2003). Room-temperature single-nucleotide polymorphism and multiallele DNA detection using fluorescent nanocrystals and microarrays. *Anal. Chem.* **75**(18), 4766–4772.
25. Zhang, C.Y., *et al.* (2005). Single-quantum-dot-based DNA nanosensor. *Nat. Mater.* **4**(11), 826–831.
26. Robelek, R., *et al.* (2004). Multiplexed hybridization detection of quantum dot-conjugated DNA sequences using surface plasmon enhanced fluorescence microscopy and spectrometry. *Anal. Chem.* **76**(20), 6160–6165.
27. Ho, Y.P., *et al.* (2005). Multiplexed hybridization detection with multicolor colocalization of quantum dot nanoprobes. *Nano. Lett.* **5**(9), 1693–1697.
28. Scholes, G.D. (2003). Long-range resonance energy transfer in molecular systems. *Ann. Rev. Phys. Chem.* **54**, 57–87.
29. Lakowicz, J.R. (1999). *Principles of Fluorescence Spectroscopy* (2nd edition). Kluwer Academic Publishers, Dordrecht, Netherlands.
30. Clapp, A.R., *et al.* (2004). Fluorescence resonance energy transfer between quantum dot donors and dye-labeled protein acceptors. *J. Am. Chem. Soc.* **126**(1), 301–310.
31. Zhou, D.J., *et al.* (2005). Fluorescence resonance energy transfer between a quantum dot donor and a dye acceptor attached to DNA. *Chem. Commun.* (38), 4807–4809.
32. Hohng, S., and Ha, T. (2005). Single-molecule quantum-dot fluorescence resonance energy transfer. *ChemPhysChem.* **6**(5), 956–960.
33. Chao, J., *et al.* (2015). DNA-based plasmonic nanostructures. *Mater. Today.* **18**(6), 326–335.
34. Mayer, K.M., and Hafner, J.H. (2011). Localized surface plasmon resonance sensors. *Chem. Rev.* **111**(6), 3828–3857.
35. Liu, Z., *et al.* (2005). Focusing surface plasmons with a plasmonic lens. *Nano. Lett.* **5**(9), 1726–1729.

36. Kuzyk, A., *et al.* (2012). DNA-based self-assembly of chiral plasmonic nanostructures with tailored optical response. *Nature.* **483**(7389), 311–314.
37. Zhou, C., Duan, X., and Liu, N. (2015). A plasmonic nanorod that walks on DNA origami. *Nat. Commun.* **6**, 8102.
38. Gur, F.N., *et al.* (2016). Toward self-assembled plasmonic devices: High-yield arrangement of gold nanoparticles on DNA origami templates. *ACS Nano.* **10**(5), 5374–5382.
39. Copp, S.M., *et al.* (2016). Cluster plasmonics: Dielectric and shape effects on DNA-stabilized silver clusters. *Nano. Lett.* **16**(6), 3594–3599.
40. Acuna, G.P., *et al.* (2012). Fluorescence enhancement at docking sites of DNA-directed self-assembled nanoantennas. *Science.* **338**(6106), 506–510.
41. Thacker, V.V., *et al.* (2014). DNA origami based assembly of gold nanoparticle dimers for surface-enhanced Raman scattering. *Nat. Commun.* **5**, 3448.
42. Sinha, S.S., *et al.* (2016). Multimodal nonlinear optical imaging of live cells using plasmon-coupled DNA-mediated gold nanoprism assembly. *J. Phys. Chem. C.* **120**(8), 4546–4555.
43. Liu, L., and Wu, H.C. (2016). DNA-based nanopore sensing. *Angew. Chem. Int. Ed. Engl.* **55**(49), 15216–15222.
44. Kasianowicz, J.J., *et al.* (1996). Characterization of individual polynucleotide molecules using a membrane channel. *Proc. Nat. Acad. Sci. USA.* **93**(24), 13770–13773.
45. Bell, N.A., *et al.* (2012). DNA origami nanopores. *Nano. Lett.* **12**(1), 512–517.
46. Hernandez-Ainsa, S., *et al.* (2013). DNA origami nanopores for controlling DNA translocation. *ACS Nano.* **7**(7), 6024–6030.
47. Kong, J., Bell, N.A., and Keyser, U.F. (2016). Quantifying nanomolar protein concentrations using designed DNA carriers and solid-state nanopores. *Nano. Lett.* **16**(6), 3557–3562.
48. Langecker, M., *et al.* (2012). Synthetic lipid membrane channels formed by designed DNA nanostructures. *Science.* **338**(6109), 932–936.
49. Nilsson, M., *et al.* (1994). Padlock probes: Circularizing oligonucleotides for localized DNA detection. *Science.* **265**(5181), 2085–2088.
50. Demidov, V.V., and Frank-Kamenetskii, M.D. (2004). Two sides of the coin: Affinity and specificity of nucleic acid interactions. *Trends Biochem. Sci.* **29**(2), 62–71.

51. Kuhn, H., Demidov, V.V., and Frank-Kamenetskii, M.D. (2002). Rolling-circle amplification under topological constraints. *Nucleic Acids Res.* **30**(2), 574–580.

52. Landegren, U., *et al.* (1988). A ligase-mediated gene detection technique. *Science.* **241**(4869), 1077–1080.

53. Nilsson, M., *et al.* (1997). Padlock probes reveal single-nucleotide differences, parent of origin and in situ distribution of centromeric sequences in human chromosomes 13 and 21. *Nat. Genet.* **16**(3), 252–255.

54. Lizardi, P.M., *et al.* (1998). Mutation detection and single-molecule counting using isothermal rolling-circle amplification. *Nat. Genet.* **19**(3), 225–232.

55. Christian, A.T., *et al.* (2001). Detection of DNA point mutations and mRNA expression levels by rolling circle amplification in individual cells. *Proc. Natl. Acad. Sci. USA.* **98**(25), 14238–14243.

56. Jahangir Tafrechi, R.S., *et al.* (2007). Single-cell A3243G mitochondrial DNA mutation load assays for segregation analysis. *J. Histochem. Cytochem.* **55**(11), 1159–1166.

57. Melin, J., *et al.* (2008). Ligation-based molecular tools for lab-on-a-chip devices. *N. Biotechnol.* **25**(1), 42–48.

58. Dahl, F., *et al.* (2004). Circle-to-circle amplification for precise and sensitive DNA analysis. *Proc. Natl. Acad. Sci. USA.* **101**(13), 4548–4553.

59. Hardenbol, P., *et al.* (2003). Multiplexed genotyping with sequence-tagged molecular inversion probes. *Nat. Biotechnol.* **21**(6), 673–678.

60. Baner, J., *et al.* (2003). Parallel gene analysis with allele-specific padlock probes and tag microarrays. *Nucleic Acids Res.* **31**(17), e103.

61. Akhras, M.S., *et al.* (2007). Connector inversion probe technology: A powerful one-primer multiplex DNA amplification system for numerous scientific applications. *PLoS One.* **2**(9), e915.

62. Stefan, C.P., Koehler, J.W., and Minogue, T.D. (2016). Targeted next-generation sequencing for the detection of ciprofloxacin resistance markers using molecular inversion probes. *Sci. Rep.* **6**, 25904.

63. Merkiene, E., *et al.* (2010). Direct detection of RNA *in vitro* and *in situ* by target-primed RCA: The impact of *E. coli* RNase III on the detection efficiency of RNA sequences distanced far from the 3′-end. *RNA.* **16**(8), 1508–1515.

64. Lagunavicius, A., *et al.* (2009). Novel application of Phi29 DNA polymerase: RNA detection and analysis *in vitro* and *in situ* by target RNA-primed RCA. *RNA.* **15**(5), 765–771.

65. Millard, P.J., *et al.* (2006). Detection of infectious haematopoietic necrosis virus and infectious salmon anaemia virus by molecular padlock amplification. *J. Fish. Dis.* **29**(4), 201–203.

66. Cheng, Y., *et al.* (2009). Highly sensitive determination of microRNA using target-primed and branched rolling-circle amplification. *Angew. Chem. Int. Ed. Engl.* **48**(18), 3268–3272.

67. Mashimo, Y., *et al.* (2011). Detection of small RNA molecules by a combination of branched rolling circle amplification and bioluminescent pyrophosphate assay. *Anal. Bioanal. Chem.* **401**(1), 221–227.

68. Nilsson, M., *et al.* (2001). RNA-templated DNA ligation for transcript analysis. *Nucleic Acids Res.* **29**(2), 578–581.

69. Lu, J., *et al.* (2004). Unique ligation properties of eukaryotic NAD+-dependent DNA ligase from Melanoplus sanguinipes entomopoxvirus. *Biochim. Biophys. Acta.* **1701**(1–2), 37–48.

70. Nilsson, M., *et al.* (2000). Enhanced detection and distinction of RNA by enzymatic probe ligation. *Nat. Biotechnol.* **18**(7), 791–793.

71. Kleppe, K., Van de Sande, J.H., and Khorana, H.G. (1970). Polynucleotide ligase-catalyzed joining of deoxyribo-oligonucleotides on ribopolynucleotide templates and of ribo-oligonucleotides on deoxyribopolynucleotide templates. *Proc. Natl. Acad. Sci. USA.* **67**(1), 68–73.

72. Fareed, G.C., Wilt, E.M., and Richardson, C.C. (1971). Enzymatic breakage and joining of deoxyribonucleic acid. 8. Hybrids of ribo- and deoxyribonucleotide homopolymers as substrates for polynucleotide ligase of bacteriophage T4. *J. Biol. Chem.* **246**(4), 925–932.

73. Takahashi, H., *et al.* (2010). Direct detection of green fluorescent protein messenger RNA expressed in Escherichia coli by rolling circle amplification. *Anal. Biochem.* **401**(2), 242–249.

74. Li, N., *et al.* (2009). Stand-alone rolling circle amplification combined with capillary electrophoresis for specific detection of small RNA. *Anal. Chem.* **81**(12), 4906–4913.

75. Stougaard, M., *et al.* (2007). *In situ* detection of non-polyadenylated RNA molecules using turtle probes and target primed rolling circle PRINS. *BMC Biotechnol.* **7**, 69.

76. Vagner, J., Steiniche, T., and Stougaard, M. (2015). *In-situ* hybridization-based quantification of hTR: A possible biomarker in malignant melanoma. *Histopathology.* **66**(5), 747–751.

77. Larsson, C., *et al.* (2010). *In situ* detection and genotyping of individual mRNA molecules. *Nat. Meth.* **7**(5), 395–397.

78. de la Pena, M., and Garcia-Robles, I. (2010). Ubiquitous presence of the hammerhead ribozyme motif along the tree of life. *RNA.* **16**(10), 1943–1950.

79. Prody, G.A., *et al.* (1986). Autolytic processing of dimeric plant virus satellite RNA. *Science.* **231**(4745), 1577–1580.

80. Hutchins, C.J., *et al.* (1986). Self-cleavage of plus and minus RNA transcripts of avocado sunblotch viroid. *Nucleic Acids Res.* **14**(9), 3627–3640.

81. de la Pena, M., and Garcia-Robles, I. (2010). Intronic hammerhead ribozymes are ultraconserved in the human genome. *EMBO Rep.* **11**(9), 711–716.

82. Breaker, R.R., and Joyce, G.F. (1994). A DNA enzyme that cleaves RNA. *Chem. Biol.* **1**(4), 223–229.

83. Breaker, R.R., and Joyce, G.F. (2014). The expanding view of RNA and DNA function. *Chem. Biol.* **21**(9), 1059–1065.

84. Achenbach, J.C., Nutiu, R., and Li, Y.F. (2005). Structure-switching allosteric deoxyribozymes. *Anal. Chim. Acta.* **534**(1), 41–51.

85. Santoro, S.W., and Joyce, G.F., (1997). A general purpose RNA-cleaving DNA enzyme. *Proc. Natl. Acad. Sci. USA.* **94**(9), 4262–4266.

86. Travascio, P., Li, Y., and Sen, D. (1998). DNA-enhanced peroxidase activity of a DNA-aptamer-hemin complex. *Chem. Biol.* **5**(9), 505–517.

87. Travascio, P., *et al.* (1999). A ribozyme and a catalytic DNA with peroxidase activity: Active sites versus cofactor-binding sites. *Chem. Biol.* **6**(11), 779–787.

88. Sun, C., *et al.* (2010). Electrochemical DNA biosensor based on proximity-dependent DNA ligation assays with DNAzyme amplification of hairpin substrate signal. *Biosens. Bioelectron.* **25**(11), 2483–2489.

89. Xuan, F., Fan, T.W., and Hsing, I.M. (2015). Electrochemical interrogation of kinetically-controlled dendritic DNA/PNA assembly for immobilization-free and enzyme-free nucleic acids sensing. *ACS Nano.* **9**(5), 5027–5033.

90. Mei, S.H., *et al.* (2003). An efficient RNA-cleaving DNA enzyme that synchronizes catalysis with fluorescence signaling. *J. Am. Chem. Soc.* **125**(2), 412–420.

91. Xu, J., *et al.* (2016). Double-stem hairpin probe and ultrasensitive colorimetric detection of cancer-related nucleic acids. *Theranostics.* **6**(3), 318–327.

92. Ellington, A.D., and Szostak, J.W. (1990). *In vitro* selection of RNA molecules that bind specific ligands. *Nature.* **346**(6287), 818–822.

93. Hermann, T., and Patel, D.J. (2000). Adaptive recognition by nucleic acid aptamers. *Science*. **287**(5454), 820–825.
94. Sefah, K., *et al.* (2009). Nucleic acid aptamers for biosensors and bio-analytical applications. *Analyst*. **134**(9), 1765–1775.
95. Jia, W., *et al.* (2016). CD109 is identified as a potential nasopharyngeal carcinoma biomarker using aptamer selected by cell-SELEX. *Oncotarget*. **7**(34), 55328–55342.
96. Cho, E.J., *et al.* (2005). Using a deoxyribozyme ligase and rolling circle amplification to detect a non-nucleic acid analyte, ATP. *J. Am. Chem. Soc.* **127**(7), 2022–2023.
97. Zhou, L., *et al.* (2007). Aptamer-based rolling circle amplification: A platform for electrochemical detection of protein. *Anal. Chem.* **79**(19), 7492–7500.
98. Huang, Y., *et al.* (2008). Electrochemical immunosensor of platelet-derived growth factor with aptamer-primed polymerase amplification. *Anal. Biochem.* **382**(1), 16–22.
99. Di Giusto, D.A., *et al.* (2005). Proximity extension of circular DNA aptamers with real-time protein detection. *Nucleic Acids Res.* **33**(6), e64.
100. Tang, L., *et al.* (2012). Colorimetric and ultrasensitive bioassay based on a dual-amplification system using aptamer and DNAzyme. *Anal. Chem.* **84**(11), 4711–4717.
101. Guo, L., Hao, L., and Zhao, Q. (2016). An aptamer assay using rolling circle amplification coupled with thrombin catalysis for protein detection. *Anal. Bioanal. Chem.* **408**(17), 4715–4722.
102. Civit, L., *et al.* (2016). Sensitive detection of cancer cells using light-mediated apta-PCR. *Methods*. **97**, 104–109.
103. Pinto, A., *et al.* (2016). Apta-PCR. *Methods Mol. Biol.* **1380**, 171–177.
104. Fredriksson, S., *et al.* (2002). Protein detection using proximity-dependent DNA ligation assays. *Nat. Biotechnol.* **20**(5), 473–477.
105. Lundberg, M., *et al.* (2011). Multiplexed homogeneous proximity ligation assays for high-throughput protein biomarker research in serological material. *Mol. Cell. Proteomics*. **10**(4), M110.004978.
106. Gustafsdottir, S.M., *et al.* (2005). Proximity ligation assays for sensitive and specific protein analyses. *Anal. Biochem.* **345**(1), 2–9.
107. Greenwood, C., *et al.* (2015). Proximity assays for sensitive quantification of proteins. *Biomol. Detect. Quantif.* **4**, 10–16.
108. Zheng, D., Zou, R., Lou, X. (2012). Label-free fluorescent detection of ions, proteins, and small molecules using structure-switching aptamers, SYBR Gold, and exonuclease I. *Anal. Chem.* **84**(8), 3554–3560.

109. Xue, L., Zhou, X., Xing, D. (2012). Sensitive and homogeneous protein detection based on target-triggered aptamer hairpin switch and nicking enzyme assisted fluorescence signal amplification. *Anal. Chem.* **84**(8), 3507–3513.
110. Yang, L., *et al.* (2007). Real-time rolling circle amplification for protein detection. *Anal. Chem.* **79**(9), 3320–3329.
111. Wu, Z.S., *et al.* (2010). Universal aptameric system for highly sensitive detection of protein based on structure-switching-triggered rolling circle amplification. *Anal. Chem.* **82**(6), 2221–2227.
112. Stougaard, M., *et al.* (2011). Strategies for highly sensitive biomarker detection by rolling circle amplification of signals from nucleic acid composed sensors. *Integr. Biol.* **3**(10), 982–992.
113. Knudsen, B.R., Jepsen, M.L., and Ho, Y.P. (2013). Quantum dot-based nanosensors for diagnosis via enzyme activity measurement. *Expert Rev. Mol. Diagn.* **13**(4), 367–375.
114. Hede, M.S., Fjelstrup, S., and Knudsen, B.R. (2015). DNA Sensors for Malaria Diagnosis. *Nano LIFE.* **5**(2), 1541003.
115. Song, Y., *et al.* (2014). Nicking enzyme-assisted biosensor for Salmonella enteritidis detection based on fluorescence resonance energy transfer. *Biosens. Bioelectron.* **55**, 400–404.
116. Zhao, G., *et al.* (2013). Enzymatic cleavage of type II restriction endonucleases on the 2′-O-methyl nucleotide and phosphorothioate substituted DNA. *PLoS One.* **8**(11), e79415.
117. Chen, Y., Wang, L., and Jiang, W. (2012). Micrococcal nuclease detection based on peptide-bridged energy transfer between quantum dots and dye-labeled DNA. *Talanta.* **97**, 533–538.
118. Deng, J., *et al.* (2012). Sensitive detection of endonuclease activity and inhibition using gold nanorods. *Biosens. Bioelectron.* **34**(1), 144–150.
119. Niu, S., *et al.* (2010). Nicking endonuclease and target recycles signal amplification assisted quantum dots for fluorescence detection of DNA. *Anal. Chim. Acta.* **680**(1–2), 54–58.
120. Qiu, T., *et al.* (2010). A positively charged QDs-based FRET probe for micrococcal nuclease detection. *Analyst.* **135**(9), 2394–2399.
121. Rutkauskas, D., *et al.* (2014). Restriction enzyme Ecl18kI-induced DNA looping dynamics by single-molecule FRET. *J. Phys. Chem. B.* **118**(29), 8575–8582.
122. Zhao, G., *et al.* (2014). Effects of cations on small fragment of DNA polymerase I using a novel FRET assay. *Acta. Bioch. Biophy. Sin.* **46**(8), 659–667.

123. Jensen, P.W., *et al.* (2013). Real-time detection of TDP1 activity using a fluorophore-quencher coupled DNA-biosensor. *Biosens. Bioelectron.* **48**, 230–237.

124. Jakobsen, A.K., *et al.* (2015). Correlation between topoisomerase I and tyrosyl-DNA phosphodiesterase 1 activities in non-small cell lung cancer tissue. *Exp. Mol. Pathol.* **99**(1), 56–64.

125. Wang, L.J., *et al.* (2016). Base-excision-repair-induced construction of a single quantum-dot-based sensor for sensitive detection of DNA glycosylase activity. *Anal. Chem.* **88**(15), 7523–7529.

126. Svilar, D., Vens, C., and Sobol, R.W. (2012). Quantitative, real-time analysis of base excision repair activity in cell lysates utilizing lesion-specific molecular beacons. *J. Vis. Exp.* (66), e4168.

127. Marcussen, L.B., *et al.* (2013). DNA-based sensor for real-time measurement of the enzymatic activity of human topoisomerase I. *Sensors.* **13**(4), 4017–4028.

128. Jepsen, M.L., *et al.* (2014). Quantum dot based DNA nanosensors for amplification-free detection of human topoisomerase I. *RSC Adv.* **4**(5), 2491–2494.

129. Kristoffersen, E.L., *et al.* (2015). Real-time investigation of human topoisomerase I reaction kinetics using an optical sensor: A fast method for drug screening and determination of active enzyme concentrations. *Nanoscale* **7**(21), 9825–9834.

130. Jepsen, M.L., *et al.* (2016). Specific detection of the cleavage activity of mycobacterial enzymes using a quantum dot based DNA nanosensor. *Nanoscale.* **8**(1), 358–364.

131. Zuccaro, L., *et al.* (2015). Real-time label-free direct electronic monitoring of topoisomerase enzyme binding kinetics on graphene. *ACS Nano.* **9**(11), 11166–11176.

132. Givskov, A., *et al.* (2016). Optimized detection of Plasmodium falciparum topoisomerase I enzyme activity in a complex Biological sample by the use of molecular beacons. *Sensors.* **16**(11), 1916.

133. Jakobsen, A.-K., and Stougaard, M. (2015). Combining a nanosensor and ELISA for measurement of tyrosyl-DNA phosphodiesterase 1 (TDP1) activity and protein amount in cell and tissue extract. *Nano LIFE.* **5**(2), 1541001.

134. Scharf, S.J., Horn, G.T., and Erlich, H.A. (1986). Direct cloning and sequence analysis of enzymatically amplified genomic sequences. *Science.* **233**(4768), 1076–1078.

135. Dhama, K., *et al.* (2014). Loop-mediated isothermal amplification of DNA (LAMP): A new diagnostic tool lights the world of diagnosis of animal and human pathogens: A review. *Pak. J. Biol. Sci.* **17**(2), 151–166.
136. Kim, J., and Easley, C.J. (2011). Isothermal DNA amplification in bioanalysis: Strategies and applications. *Bioanalysis* **3**(2), 227–239.
137. Yan, L., *et al.* (2014). Isothermal amplified detection of DNA and RNA. *Mol. Biosyst.* **10**(5), 970–1003.
138. Mizuta, R., Mizuta, M., and Kitamura, D. (2003). Atomic force microscopy analysis of rolling circle amplification of plasmid DNA. *Arch. Histol. Cytol.* **66**(2), 175–181.
139. Beyer, S., Nickels, P., and Simmel, F.C. (2005). Periodic DNA nanotemplates synthesized by rolling circle amplification. *Nano. Lett.* **5**(4), 719–722.
140. Yan, J., *et al.* (2010). An on-nanoparticle rolling-circle amplification platform for ultrasensitive protein detection in biological fluids. *Small.* **6**(22), 2520–2525.
141. Cheglakov, Z., *et al.* (2007). Diagnosing viruses by the rolling circle amplified synthesis of DNAzymes. *Org. Biomol. Chem.* **5**(2), 223–225.
142. Akhtar, S., *et al.* (2010). Real-space transmission electron microscopy investigations of attachment of functionalized magnetic nanoparticles to DNA-coils acting as a biosensor. *J. Phys. Chem. B.* **114**(41), 13255–13262.
143. Cheng, W., *et al.* (2010). Cascade signal amplification strategy for subattomolar protein detection by rolling circle amplification and quantum dots tagging. *Anal. Chem.* **82**(8), 3337–3342.
144. Zhu, G., *et al.* (2013). Noncanonical self-assembly of multifunctional DNA nanoflowers for biomedical applications. *J. Am. Chem. Soc.* **135**(44), 16438–16445.
145. Stougaard, M., *et al.* (2009). Single-molecule detection of human topoisomerase I cleavage-ligation activity. *ACS Nano.* **3**(1), 223–233.
146. Andersen, F.F., *et al.* (2009). Multiplexed detection of site specific recombinase and DNA topoisomerase activities at the single molecule level. *ACS Nano.* **3**(12), 4043–4054.
147. Baner, J., *et al.* (1998). Signal amplification of padlock probes by rolling circle replication. *Nucleic Acids Res.* **26**(22), 5073–5078.
148. Juul, S., *et al.* (2012). Droplet microfluidics platform for highly sensitive and quantitative detection of malaria-causing Plasmodium parasites based on enzyme activity measurement. *ACS Nano.* **6**(12), 10676–10683.
149. Nallur, G., *et al.* (2001). Signal amplification by rolling circle amplification on DNA microarrays. *Nucleic Acids Res.* **29**(23), E118.

150. Wang, J., *et al.* (2016). Novel DNA sensor system for highly sensitive and quantitative retrovirus detection using virus encoded integrase as a biomarker. *Nanoscale.* **9**(1), 440–448.
151. Juul, S., *et al.* (2011). Detection of single enzymatic events in rare or single cells using microfluidics. *ACS Nano.* **5**(10), 8305–8310.
152. Kristoffersen, E.L., *et al.* (2015). Molecular beacon enables combination of highly processive and highly sensitive rolling circle amplification readouts for detection of DNA-modifying enzymes. *Nano LIFE.* **5**(2), 1541002.
153. Wang, J.C. (2002). Cellular roles of DNA topoisomerases: A molecular perspective. *Nat. Rev. Mol. Cell. Biol.* **3**(6), 430–440.
154. Leppard, J.B., and Champoux, J.J. (2005). Human DNA topoisomerase I: relaxation, roles, and damage control. *Chromosoma.* **114**(2), 75–85.
155. Pommier, Y. (2006). Topoisomerase I inhibitors: Camptothecins and beyond. *Nat. Rev. Cancer.* **6**(10), 789–802.
156. Tesauro, C., *et al.* (2012). Specific detection of topoisomerase I from the malaria causing P. falciparum parasite using isothermal rolling circle amplification. *Conf. Proc. IEEE Eng. Med. Biol. Soc.* 2416–2419.
157. Waldminghaus, T., *et al.* (2005). RNA thermometers are common in alpha- and gamma-proteobacteria. *Biol. Chem.* **386**(12), 1279–1286.
158. Tyagi, S., and Kramer, F.R. (1996). Molecular beacons: Probes that fluoresce upon hybridization. *Nat. Biotechnol.* **14**(3), 303–308.
159. Jonstrup, A.T., Fredsoe, J., and Andersen, A.H. (2013). DNA hairpins as temperature switches, thermometers and ionic detectors. *Sensors.* **13**(5), 5937–5944.
160. Ke, G.L., *et al.* (2012). L-DNA molecular beacon: A safe, stable, and accurate intracellular nano-thermometer for temperature sensing in living cells. *J. Am. Chem. Soc.* **134**(46), 18908–18911.
161. Lal, S., Clare, S.E., and Halas, N.J. (2008). Nanoshell-enabled photothermal cancer therapy: Impending clinical impact. *Accounts Chem. Res.* **41**(12), 1842–1851.
162. Huang, X., *et al.* (2011). Freestanding palladium nanosheets with plasmonic and catalytic properties. *Nat. Nanotechnol.* **6**(1), 28–32.
163. Han, J., and Burgess, K. (2010). Fluorescent indicators for intracellular pH. *Chem. Rev.* **110**(5), 2709–2728.
164. Scopes, R.K. (2001). *Enzyme Activity and Assays, in eLS.* John Wiley & Sons, Ltd., New Jersey, USA.
165. Halder, S., and Krishnan, Y. (2015). Design of ultrasensitive DNA-based fluorescent pH sensitive nanodevices. *Nanoscale.* **7**(22), 10008–10012.

166. Lannes, L., *et al.* (2015). Tuning the pH Response of i-Motif DNA Oligonucleotides. *Chembiochem.* **16**(11), 1647–1656.
167. Modi, S., *et al.* (2013). Two DNA nanomachines map pH changes along intersecting endocytic pathways inside the same cell. *Nat. Nanotechnol.* **8**(6), 459–467.
168. Modi, S., and Krishnan, Y. (2011). A method to map spatiotemporal pH changes inside living cells using a pH-triggered DNA nanoswitch. *Meth. Mol. Biol.* **749**, 61–77.
169. Modi, S., *et al.* (2014). Recombinant antibody mediated delivery of organelle-specific DNA pH sensors along endocytic pathways. *Nanoscale* **6**(2), 1144–1152.
170. Amodio, A., *et al.* (2014). Rational design of pH-controlled DNA strand displacement. *J. Am. Chem. Soc.* **136**(47), 16469–16472.
171. Saha, S., Chakraborty, K., and Krishnan, Y. (2012). Tunable, colorimetric DNA-based pH sensors mediated by A-motif formation. *Chem. Commun.* **48**(19), 2513–2515.
172. Thomas, J.M., Yu, H.Z., and Sen, D. (2012). A mechano-electronic DNA switch. *J. Am. Chem. Soc.* **134**(33), 13738–13748.
173. Mao, C., *et al.* (1999). A nanomechanical device based on the B–Z transition of DNA. *Nature.* **397**(6715), 144–146.
174. Hu, W.T., Trojanowski, J.Q., and Shaw, L.M. (2011). Biomarkers in frontotemporal lobar degenerations — progress and challenges. *Prog. Neurobiol.* **95**(4), 636–648.
175. Lari, S.A., and Kuerer, H.M. (2011). Biological markers in DCIS and risk of breast recurrence: A systematic review. *J. Cancer.* **2**, 232–261.
176. Doherty, T.M., Wallis, R.S., and Zumla, A. (2009). Biomarkers of disease activity, cure, and relapse in tuberculosis. *Clin. Chest. Med.* **30**(4), 783–796.
177. Vitzthum, F., *et al.* (2005). Proteomics: From basic research to diagnostic application. A review of requirements & needs. *J. Proteome Res.* **4**(4), 1086–1097.
178. Barbee, K.D., Chandrangsu, M., and Huang, X. (2011). Fabrication of DNA polymer brush arrays by destructive micropatterning and rolling-circle amplification. *Macromol. Biosci.* **11**(5), 607–617.
179. Banerjee, A., *et al.* (2013). Controlled release of encapsulated cargo from a DNA icosahedron using a chemical trigger. *Angew. Chem. Int. Ed. Engl.* **52**(27), 6854–6857.

Chapter 4

DNA Nanostructures in Cell Biology and Medicine

Liqian Wang*,‡ and Giuseppe Arrabito†

Laboratory of Physical Biology
Shanghai Institute of Applied Physics
Chinese Academy of Science, Shanghai 201800, China
†*Department of Physics and Chemistry*
University of Palermo, Viale delle Scienze, Parco d'Orleans II
90128 Palermo, Italy
‡*wangliqian@sinap.ac.cn*

This chapter gives a clear description of DNA nanostructures in physiological environments and the mechanism of them entering cells. This chapter also summarized the most advanced DNA nanostructures for drug delivery and other applications, such as cell- systematic evolution of ligands by exponential enrichment (SELEX), dip-pen nanolithography (DPN) and, etc.

4.1 Introduction

DNA nanotechnology has enabled a new idea of building smart therapeutic nanodevices and drug delivery systems which has several advantages over conventional ones. In conventional nanomedicine, people tend to use inorganic materials, which are toxic and difficult to degrade in human body [1–3]. On the other hand, artificial DNA nanostructures are biocompatible, highly stable and programmable. As the fact that DNA is existed in human cells, the cells have undergone millions of years of evolution to learn to use, to process and to synthesis DNA. This special bond between DNA and cells makes DNA an ideal material in cell biology and medicine. DNA nanostructures also has other unique properties, such as precise structural assemblage at 1D, 2D and 3D. Previous studies have suggested the optimal size of a drug carrier is between 20 and 100 nm [4–6]. When nanoparticle reaches more than 20 nm, it can avoid renal clearance and enhance drug delivery, thus the high loading capacity of DNA structural based nanocarriers can escape renal clearance and also they are small enough to get into tumor regions [7]. DNA nanostructures are noticed to be an ideal drug delivery carrier not only because of the size, but also because DNA nanostructures can carry multiple drugs at one time and control multiple drug releases. In this chapter, we summarize the recent progress towards the idea of bringing DNA nanostructures into cells; we discuss the stability of DNA nanostructures in cellular environment and most advanced DNA nanocarriers for drug delivery. We begin with how DNA aptamer recognize live cells and then manipulate single cell by DNA–protein immobilization.

4.2 Cell-SELEX for Cellular Detection

The ability to detect and understand changes in cell conditions at a molecular level is of fundamental importance for the accurate diagnosis and timely therapy of diseases. As previously discussed in Chapter 2, DNA aptamers are excellent candidates as molecular probes able to recognize extracellular matrix signatures of cancer cells and target cell-specific ligands for therapeutic purposes. The selection process for DNA aptamers able to recognize live cells is defined cell-SELEX [8].

The process begins with the preparation of a oligonucleotide library and the growth of target cell cultures. Target cells are incubated with the DNA pool, cell-bound oligonucleotides are collected in order to generate a new enriched pool through the amplification of bound oligonucleotides. Incubation of DNA pool on control cells (counter-selection) is performed in order to reduce non-specific binding. Aptamer binding affinity towards target cells is quantified by flow cytometry. The selection process of oligonucleotide sequences is repeated until the binding assay indicates enough affinity and specificity to the target cells (typically 10–20 rounds are performed). The last selected DNA pool is cloned in *Escherichia coli*, and sequenced in order to generate candidate sequences. These are selected, synthesized and applied to binding assays. DNA aptamers have to be optimized my minimizing length (in order to reduce cost) and maximizing binding affinity. Prediction of secondary structures is carried out by the program *mfold* (http://unafold.rna.albany.edu/?q=DINAMelt). By comparing binding affinities among selected DNA aptamers belonging to the same pool, critical sequences for target binding are predicted [9].

DNA aptamer selection on cells was firstly performed against whole cells [10]. At that time — i.e. 2003 — aptamers involved cancer detection were limited by the absence of aptamers targeting cancer cell membrane proteins. By 2003, few cancer biomarkers had been identified. Prof. Tan group established a new strategy of selecting aptamers specific to whole live cells. Notably, phenotypic variations between cell types, such as between normal and cancer cells give rise to differences in molecular signature. Therefore, the ability to isolate aptamers based on these molecular signatures permits to create molecular probes specific to cancer cell types for use in cancer diagnosis and therapy. This process, known as cell-based SELEX, or cell-SELEX, was first developed using a human acute lymphoblastic leukemia cell line, CCRF–CEM (T-cell line), as the target cell [11].

The result of the process was the generation of a panel of aptamers specifically binding target cells, including one able to bind to a membrane protein tyrosine kinase 7 — a biomarker for leukemia [12].

DNA aptamers have been used to build targeting DNA nanostructures, aptamers–drug conjugates and smart DNA diagnostic/therapeutic networks (see Figure 4.1) [13]. Examples of applications in intracellular

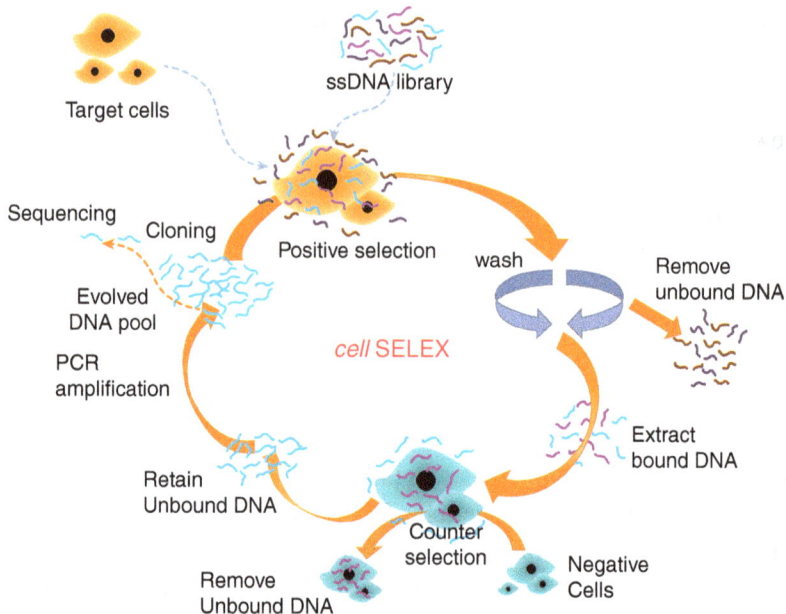

Figure 4.1. Schematic representation of cell SELEX process. Reproduced in part with permission of The Royal Society of Chemistry from Ref. [13].

biomolecular imaging by engineered DNA nanostructures containing aptamers have been shown for ATP by Fan *et al.* [14]. or mRNA by Tan *et al.* [15].

One of the most interesting applications of DNA aptamers is the detection of cancer cells, in particular circulating tumor cells, which leak from a metastasis into blood vessels, and transfer to distant locations, with the potential to establish metastatic tumors [16]. Remarkably, the sensitive detection of cancer cells plays a critical role in the early diagnosis of cancer and eventually allows for separation form benign cells. As examples of DNA aptamers integration in cell sensing devices, Qu *et al.* [17]. designed an electrochemical sensor by using two cell-specific aptamers, TLS1c and TLS11a which recognize MEAR cancer cells, with the possibility to detect as few as a single MEAR cell in 10^9 blood cells. Li *et al.* [18]. showed an aptamer functionalized hydrogel that could catch cancer cells with a density over 1000 cells/mm^2.

Interestingly, DNA aptamers have also been shown to constitute fundamental tools for novel DNA-based therapies. We here briefly discuss about integration of DNA aptamers in nanostructures based on DNA usable in chemotherapy, gene therapy and immunotherapy.

Given their high loading efficiency and cell type-specific recognition capability, DNA nanostructures can improve distribution and efficacy of chemotherapeutic drugs *in vitro* and *in vivo* [19]. Many chemotherapeutic agents have been conjugated with DNA to form covalent or non-covalent conjugation for chemotherapy (see e.g. Ref. [20]). Aptamer-integrated DNA nanostructures have been developed for targeted delivery of therapeutic drugs to reduce side effects as shown by Zhu *et al.* [21] in his aptamer-tethered DNA nanotrains (aptNTrs) for targeted chemotherapy.

Gene therapy is considered one of the most promising approach for treating diseases such as cancer or viral infections. In this therapeutic approach, small interfering RNA (siRNA) and microRNA (miRNA) molecules are employed as gene silencing tools. The bottleneck of this approach is constituted by the efficiency of gene delivery vectors. In this sense, DNA nanostructures have been explored as suitable non-viral vector candidates because of their high payload capacity and excellent biocompatibility, flexibility and mechanical stability. Examples of integration of DNA aptamers within DNA-based nanostructures usable as vectors are shown by Giangrande *et al.* [22]. who used aptamer–siRNA chimeric RNAs to deliver functional siRNAs to target cells or by Zhang *et al.* [23]. who constructed an aptamer–siRNA complex able to deliver siRNA *in vivo*, in which aptamers and siRNA are hybridized to the outermost layer of DNA dendritic nanostructures.

DNA nanostructures have been reported to conjugate with vaccine-based, cytosine-based guanine oligodeoxynucleotide (CpG)-based immunotherapies to target cancer cells. CpG motif have demonstrated to be an immunotherapeutic adjuvant in the treatment of a wide range of diseases, since it can stimulate dendritic cells, B cells, and macrophages through interaction with the toll-like receptor 9 (TLR9) [24]. For example, Liu *et al.* [25] employed a tetrahedral DNA nanostructure as a scaffold to assemble a model antigen and CpG adjuvants together into nanoscale complexes. In cell-based immunotherapy, killer lymphocytes have been used to kill cancer cells. However, these cells have to be targeted to cancer cells. DNA aptamers anchored on the cell surface of leukemia cell lines have been shown by Xiong *et al.* [26] to improve immune-efficacy of these cells and cytotoxicity towards cancer cells. Although the potentiality of DNA aptamers in cell biology has been shown, there are still some limitations that hamper applications in biomedical field such as the lengthy process of cell-SELEX and the technical difficulty in the precise identification of aptamer targets, since they are often constituted by membrane proteins.

4.3 Single Cell Manipulation by DNA-directed Protein Immobilization

One of the most intriguing challenges in modern bioanalytical sciences is the rapid and accurate identification of a large number of different molecular species with high temporal and spatial resolution [27]. This analytical capability is of outmost importance in the characterization of cellular samples at the single cell level — allowing for truly molecular quantitative analysis. In fact, any cell phenotype is determined by biomolecular network states maintained by the dynamics of multiple molecular interactions. The effect of a drug molecule is, basically, based on the perturbation of molecular interactions in such system. The capability to quantitatively analyze a single cell is then crucial for understanding the precise cellular state, especially in the case of medically relevant conditions such as necrosis, apoptosis, proliferation, differentiation, transformation, senescence, growth or shrinkage which are all associated with a profound perturbation of the cytoplasmic content.

4.3.1 *DNA-directed immobilization*

The employment of surface-based bioanalytical tools like microarrays has allowed for breakthrough in fundamental and applied biomedical research, in the form of DNA and protein microarrays. For better understanding complex cellular features in healthy or diseased states, researchers can use microarrays to delineate on a molecular level the characteristics of individual cells within complex mixtures of cells belonging to different populations. As a consequence, it is possible to quantitatively identify distinct cellular types/states according to the expression of particular cell surface molecules, and also by analyzing their response to precise signals through the secretion of factors or cellular activities. In this regard, it becomes of fundamental importance to establish an efficient immobilization strategy, which allows for oriented immobilization of small molecules and protein molecules usable for capturing cells. The main issue in this regard is constituted by the ability to retain the biological functionality of the immobilized biomolecule in order to be able to function as capture probes.

In this regard, DNA-directed immobilization (DDI) of oligonucleotide-tagged components on DNA microarrays bearing complementary oligonucleotides is a powerful and chemically mild process, which fulfils the aforementioned requirements. The DNA-directed immobilization method [28] takes advantage of specific hybridization of complementary oligonucleotides and thereby allows the site-specific capture of biomolecules by the DNA microstructures on a solid substrate (see Figure 4.2).

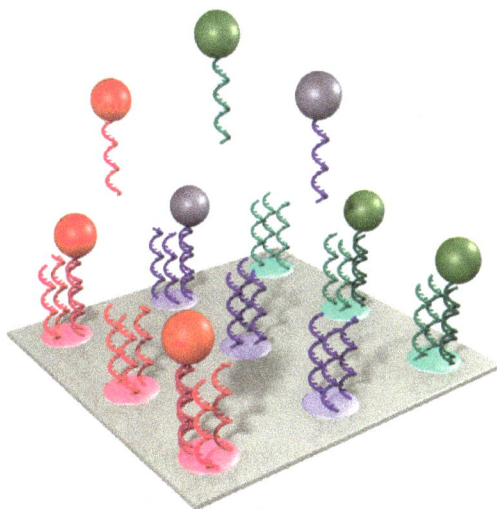

Figure 4.2. Scheme of DDI. Reprinted with permission from Ref. [28]. Copyright (2014) Elsevier.

In addition to numerous applications in the area of protein chips for diagnostics [29], the DDI method has also been employed for the generation of bioarrays of living cells [30].

Reisewitz *et al.* [31] firstly showed the possibility to build up specific cellular capture platforms by DNA directed immobilization of specific cellular antibodies, i.e. the cell specific mouse-anti human CD81 IgG (αCD81). They employed a biotinylated IgG coupling it with covalent conjugates of STV and ssDNA oligonucleotides which were hybridized with a DNA microarray containing complementary capture oligonucleotides, spotted by piezoelectrical inkjet printing on modified glass surfaces. The resulting antibody array was employed for the adhesion of HEK293 cells and Hodgkin lymphoma L540cy cells. Remarkably, the adhered cells showed viability and ability to propagate to form a densely grown monolayer, restricted to the lateral dimensions of the DNA spots inside the array. The authors showed the feasibility of their approach by employing two different set-ups, static incubation and hydrodynamic flow inside a microfluidic channel. This last approach permitted to execute cellular capture and selection of cells from culture medium. The microfluidic polymer chip consisted of PDMS substrate containing a microfluidic channel and inlet/outlet ports, mounted on a top a glass slide bearing a DNA microarray to enable IgG binding by DDI. The authors observed that it was possible to obtain optimal cellular capture at flow rates as high as 1 μL/min, whereas a higher flow rates determined significant decrease of immobilization efficiency.

Given the possibility to capture cells to DNA modified surfaces, the possibility to employ DNA arrays as capture matrix can permit the development of the single-cell biochips in which a single cell is immobilized on an array containing different microspots with lateral dimensions which are smaller than the lateral size of a single cell. This would permit quantitative analysis at the single cell level.

4.3.2 *Dip-pen nanolithography coupled with DDI for single cells interaction arrays*

Single cell microarrays require features smaller than 10 μm for functional manipulation of sub-cellular structures. Several top-down methodologies

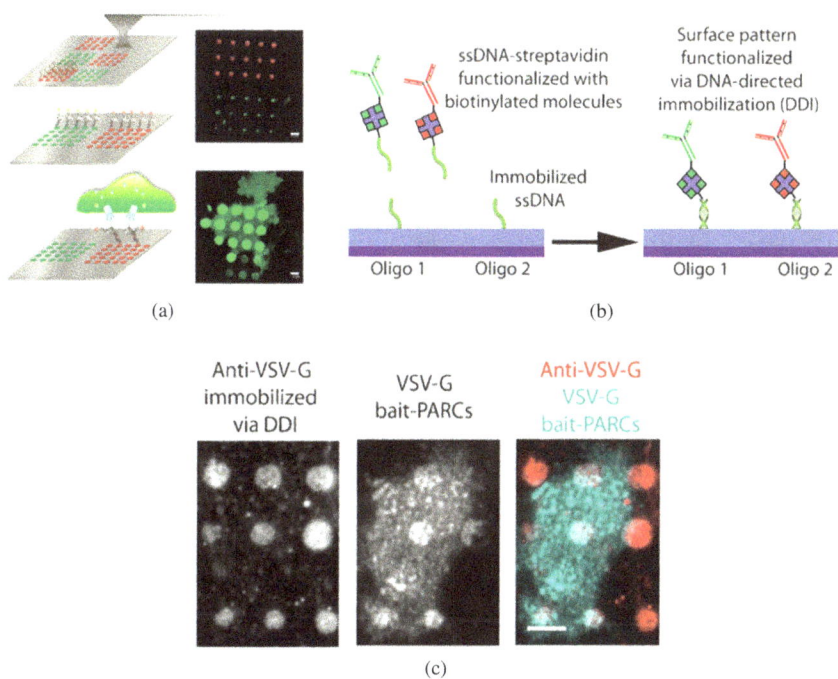

(a)

(b)

(c)

Figure 4.3. (a) Fabrication of microarrays by Dip Pen Nanolithography for DNA-directed protein immobilization and subsequent generation of live-cell arrays for the recruitment of transmembrane Epidermial growth factors (EGFR) receptors on MCF 7 cells. Reproduced from Ref. [32]. Copyright © 2013 by John Wiley and Sons, Inc. Reprinted with permission from John Wiley and Sons, Inc. (b) Scheme of generation of multiplexed antibodies microarrays by DDI. (c) Bait- PARCs displaying VSV-G epitope tags are recruited to anti-VSVG functionalized surface patterns within the plasma membrane of COS7 cells. Scale bar is equal to 5 microns. Reproduced from Ref. [33]. Copyright © 2014 by John Wiley and Sons, Inc. Reprinted with permission from John Wiley and Sons, Inc.

like electron beam lithography and microcontact printing can be employed for indirect surface patterning at this scale, however, those approaches often require clean rooms and multiplexing of several different biomolecules on the same surface is limited [27]. In this regard, Arrabito *et al.* combined DPN and DDI of proteins to fabricate cell-compatible functionalized glass surfaces (see Figure 4.3) [32].

DPN was used to realize multiplexed patterns of capture-oligonucleotides onto activated glass surfaces. In general, DPN allows for direct deposition of molecules from an ink-loaded tip onto a surface. One has to consider that the diffusive or liquid aggregation state of the ink determines deposition mechanism. In fact, in the diffusive case, molecules are dispensed to the surface through a water meniscus that forms naturally when the tip is brought in contact to the surface. Such process is dependent on molecular diffusion rate and it is reported to be slow for DNA molecules (10^{-12}–10^{-13} m^2/s). On the other hand, in the case of liquid inks, ink molecules remain homogeneously dissolved in the carrier solution. When the ink-loaded pen contacts the surface, a liquid meniscus is formed, which is composed of the liquid ink itself and water from the atmosphere. By pulling the tip away from the surface, the meniscus breaks and the spot is formed. In the liquid ink mechanism, the deposition process depends on the relative humidity (i.e. the amount of water vapor present in air expressed as a percentage of the amount needed for saturation at the same temperature) surface tension in between liquid and tip, liquid and surface as well as on the liquid carriers' viscosity. Remarkably, the deposition process is not dependent upon molecular weight of the deposited biomolecule, so it is suitable for depositing proteins and DNA molecules. On the other hand, it is dependent on the physical–chemical properties of the liquid carried in which the DNA is dissolved.

In this regard, Arrabito *et al.* optimized a strategy that employs oligonucleotide Dip-pen Nanopatterning of oligonucleotide dissolved in Polyethylene glycol (PEG) matrixes. PEG is a water-soluble and biocompatile polymer commonly used as an additive in solutions for biological applications. Complementary sequences conjugated to a protein of interest were hybridized to the spotted sequence via DDI. A systematic study on the parameters affecting printing efficiency was performed. Parameters like oligonucleotide concentration, molecular weight of PEG, dwell time, relative humidity were investigated in order to derivate the best conditions for hybridization efficiency and array writing throughput. The results of these studies was that the best conditions between small feature sizes and high density of immobilized oligonucleotide are obtained by PEG 1000 inks showing relative humidity as high as 30%, with a capture oligonucleotide concentration of 100 μM.

Oligonucleotides complementary to the immobilized capture oligonucleotides were covalently linked to streptavidin, and the resulting conjugates were functionalized with biotinylated antibodies and fluoro-phores. These streptavidin–antibody complexes then bind to the immobi-lized capture-oligonucleotide arrays. In particular, the produced arrays were functionalized with epidermal growth factor (EGF) taking advantage of covalent ssDNA–streptavidin conjugates as adaptor molecules. The surface-immobilized EGF was used for recruiting EGFR in the plasma membrane of MCF7 cells.

The combination of DPN with DNA directed immobilization has also been shown for the fabrication of multiplexed patterns of capture-oligonucleotides onto activated glass surfaces — this meaning that it was possible to generate microscale arrays of two capture oligonucleotides, corresponding after DDI to two different proteins, at dimensions below single cells [33]. This was possible by inking Dip-pen arrays with two different oligonucleotides by a microfluidic inking device. At the outset the first sub-array is written, then after a translation along the X-axis and a pens microscale alignment on the previous pattern, the double-oligonu-cleotide array is completed with the second sub-array. The pen array is translated along the X-axis of 66 μm to the right and along the Y-axis of 100 μm on to write new arrays.

This multiplexed microarray methodology was applied to simulta-neously measure the interaction of two bait-presenting artificial recep-tor constructs (PARCs) — i.e. the regulatory domain RII-b and the regulatory domain RI-a of protein kinase A (PKA) — with the prey protein construct constituted by the catalytic subunit mCherry-cat-α of PKA fused to the fluorescent protein mCherry inside individual living cells. The two different PARCs were orthogonally attached to the sur-face since in their extracellular region, they showed two different epitopes, respectively VSV-G bait and HA bait that are selectively captured by respective biotinylated antibodies linked to the ssDNA–streptavidin conjugate which is hybridized to the complementary ssDNA deposited by DPN. In Fig. 4.3, we show a single cell recruited on an anti-VSVG functionalized surface. For this, micrometer-scale arrays of antibodies with binding specificity for the peptide epitope on the bait-PARC were generated by combining DPN and DDI. It was possible to successfully discriminate the interaction of RII-b and RI-a with mCherry-cat-α [33].

4.3.3 High-throughput patterning methodologies for single cells interaction arrays

To improve the efficiency of multiplexed surface patterning, Arrabito *et al.* developed a prototype of a robust custom plotter based on 2D polymer-pen lithography (2D-PPL), a patterning methodology in which a soft elastomeric polymer (polydimethylsiloxane) delivers inks onto a surface by controlling the movement of the pen array with a scanning probe microscope [34]. A typical polymer pen array contains thousands of pyramid-shaped tips made with a master prepared by conventional photolithography and subsequent wet chemical etching. These pyramids are connected by a thin PDMS backing layer (thickness around 50–100 nm) that is adhered to a glass support [35]. This device enables rapid fabrication of microarrays at ambient conditions in a multiplexed direct-writing mode. The printing process was carried out by polymeric pyramidal pens onto which multiple (up to 36) ssDNA solutions can be loaded through a microfluidic inkwell device. Subsequent to optimization of ink viscosity and surface tension by glycerol and tween-20, DNA arrays were plotted and used for DDI of EGF-bearing ssDNA–streptavidin conjugates. The resulting microarrays covered areas of about 0.5 cm^2 using up to three different oligonucleotides. Typical feature sizes are 5 μm diameter with a pitch of 15 μm, leading to densities of up to 10^4–10^5 spots/mm^2.

The microarrays were capable of recruiting and activating EGF receptors in sub-cellular regions within human MCF7 cells. In order to verify if immobilized EGF was able to activate EGF receptors, cells were stained with antibodies specifically directed against active EGFR, which is phosphorylated at tyrosine 1068. The ratio between phosphorylated and total EGFR was used to measure the activation state of this receptor. In this sense, the authors found out that a significantly higher fraction of EGF receptors was phosphorylated within cell regions that contact EGF functionalized surfaces. Such results confirm the ability of Polymer Pen Lithography to produce functional sub-cellular scale EGF arrays able to activate EGF receptors in cells (see Figure 4.4).

In a very recent breakthrough paper, Angelin *et al.* [36]. demonstrated the site-directed sorting of differently encoded, protein-decorated DNA-origami structures on DNA microarrays. The combination of bottom-up

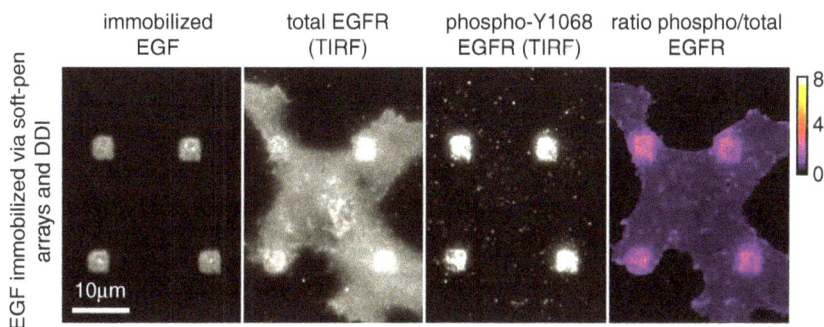

Figure 4.4. Experimental evidence of EGF Receptor activation by surface immobilized EGF ligands. EGF was immobilized on oligonucleotide arrays printed on glass substrates. To visualize successful immobilization, the DNA–STV–biotin–EGF complexes contained biotin-Atto740. MCF7 cells expressing EGFP–EGFR were cultured on those functionalized surfaces and visualized by Total internal Reflectance Fluorescence microscopy. Immunostaining using phosphor-specific revealed increased phosphorylation of EGFR at Tyrosine residue 1068 in subcellular regions contacting spots with immobilized EGF. Reproduced from Ref. [34]. Copyright © 2014 by John Wiley and Sons, Inc. Reprinted with permission from John Wiley and Sons, Inc.

self-assembly of protein–DNA nanostructures and top-down micropatterning — carried out by polymer pen lithography–of solid surfaces enabled the creation of multiscale origami structures as interface for cells (MOSAIC). They designed a rectangular origami construct from the 5438 nucleotides template 109Z5 having nine single-stranded DNA (ssDNA) binding tags, protruding from one side of the plane of the quasi-2D nanostructure. The nine binding sites are able to bind to their complementary surface-bound capture oligonucleotides.

This innovative technology was applied to investigate the activation of EGF receptors in living MCF7 cells through distinctive nanoscale arrangements of EGF ligands. This technology can represent a significant breakthrough in bioanalytical applications since it permits the assembly of structures having the same size of biomolecular assemblies present in the membrane of living cells. Such assemblies are composed of tens to thousands of molecules (sizes around 5–100 nm) and play a crucial role in the outcome of signaling events and the development of the cellular phenotype.

4.4 How to Get into Cells

To get into cells, we firstly need to understand the entries. Such as human civilization that if we want to export our goods to foreign countries we need to know the location of the ports, cultures and policies. Similar situation happens to cells, they must be accessed through entry points, and this is related to "protector" of cells — membranes. The membrane encloses the cell and maintains the essential differences between intercellular and extracellular environments. Inside cells, there are organelle membranes that do the same job to keep each organelle in different environment than cytosol. So, membranes are critical and unavoidable entrees for big or small molecules, as well as DNA nanostructures. In this part, we will discuss the different regulated modes of getting into cells.

The entry processes for macromolecules (proteins and DNA nanostructures) and small molecules (ions and amino acids) to get into cells are different. For macromolecules, they are transported into cells under a number of active and regulated processes that is called endocytosis (endo "within" cytosis "cell") [7, 37]. It is a process in which a macromolecule gets into a cell without passing through the cell membrane. Usually, the endocytosis process falls into two categories: phagocytosis and pinocytosis (Figure 4.5). These two categories can be mainly distinguished by the size of the object ingested. Mammalian phagocytosis is limited to particular cells, such as macrophages, monocytes, dendritic cells and neutrophils; however, pinocytosis happens to all cells through four endocytic uptake pathways: macro-pinocytosis, clathrin-mediated endocytosis (CME), caveolae-mediated endocytosis, and clathrin/caveolin-independent endocytosis. The relationships among those processes are shown in Figure 4.5.

4.4.1 *Phagocytosis*

Phagocytosis, which means "cell-eating," is the process in which cells swallow large objects. As shown in Figure 4.5, during phagocytosis, the membrane first folds around the target, and then the object are sealed into

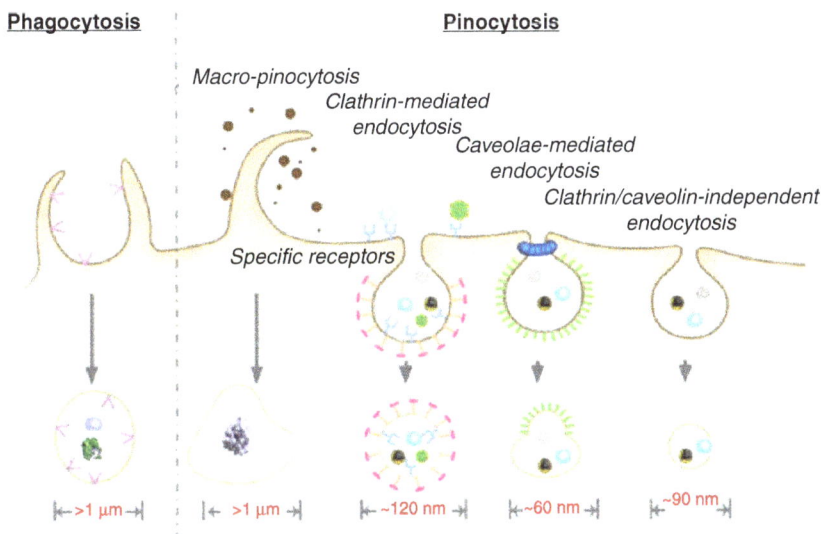

Figure 4.5. Multiple gateways into living cells. Different endocytic mechanisms use different sized endocytic vesicles to bring in a variety of cargoes.

large vesicles. The large vesicles usually called as phagosome, which is generally larger than 250 nm in diameter. Usually, the defense organisms against non-self intruders, such as bacteria, viruses and drug delivery nanostructures, trigger phagocytosis [38]. The process starts with the recognition of target particles and the non-adhesion of opsonized particles to the macrophage, and then the macrophages ingest the particles [39]. During the process, opsonized particles are connected to the macrophage through specific receptor-ligand interactions, which will trigger actin assembly and swallow-up of the target particles. Ultimately, the phagosome fuses with the lysosome to form phagolysosome. This is where the degradation of endocytosed particle goes. There is one directly properties of ingested particles that affect the rate of phagocytosis: size. Usually, when particles are smaller than 250 nm in diameter, they are less effectively internalized [40]. The size of the particles in serum always affect the extent of opsonization; therefore, influence the rate of phagocytosis [41].

4.4.2 *Pinocytosis*

Pinocytosis is another category of endocytosis, which literally means "cell-drinking." Different from phagocytosis, this process involves the uptake of fluids and using small vesicles. As mentioned before, this process can be split to four uptake pathways:

4.4.3 *Macro-pinocytosis*

Macro-pinocytosis is similar to phagocytosis, it takes big particles and it is driven by actin. The difference is the actin-driven membrane protrusions do not fold around the targets but collapse onto and fuse with the plasma membrane and then forming large macropinosomes with sizes comparatively to phagosomes (Figure 4.6(a)). The intracellular destiny of macropinosomes varies with cell type, but most commonly these macropinosomes will fuse with lysosomes and be observed as acidification and shrinkage [42].

4.4.4 *Clathrin-mediated endocytosis*

CME is claimed as the major route for endocytosis in most cells. CME can be divided into two types: receptor-dependent and receptor-independent. Both types are also internalizing target particles and digesting for degradative lysosomes. Receptor-dependent CME is a very common pathway for ligand–receptor complexes to get through [43], thus it is also a very important pathway for drug delivery carriers that aim at these receptors. CME happens in the membrane region that is clathrin-enriched. Clathrin is mainly cytosolic coat protein, during CME it deforms the membrane into a coated pit and then leading to a ~120 nm acidified early endosome which ultimately will fuse with a prelysosomal vesicle containing enzymes leading to a late endosome. The late endosome will turns into a lysosome at the end [44] (Figure 4.6(b)). If the target is aiming at the cytosol, our drug carriers must escape from the endosome fast enough to avoid lysosomal degradation. The other type of CME, the receptor-independent CME, has a slower internalization rate compares to the other type [45]. During this type of process, the particles usually interact with the plasma membrane via non-specific charges or hydrophobic interactions. However, the destiny goes the same way as receptor-dependent CME.

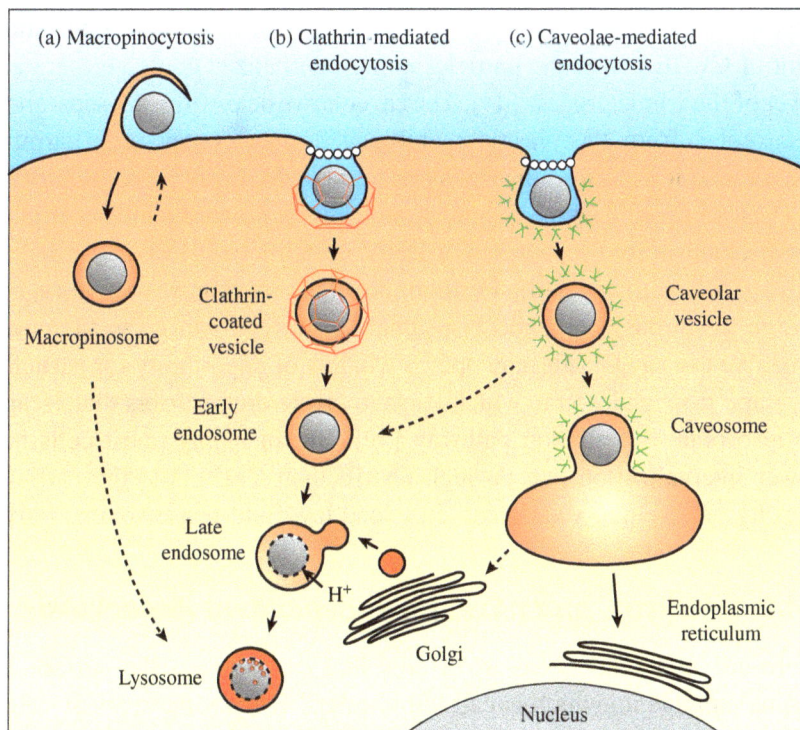

Figure 4.6. Diverse intracellular fates of carriers after different endocytic uptakes. (a) Macropinocytosis forms macropinosomes of large cargos, which are subsequently fused with lysosomes. (b) CME gives rise to an early endosome, which is acidic, followed by a late endosome after fusing with pre-lysosomal vesicles containing enzymes (red vesicle), and finally coalesces into acidic lysosomes, which contain high levels of degradative enzymes. (c) Caveolae-mediated endocytosis leads to the formation of a caveosome, possibly bypassing enzymatic degradation of the endo-lysosomal pathway. Reproduced with permission from Springer from Ref. [42].

4.4.5 *Caveolae-mediated endocytosis*

Caveolae are small (~60 nm) flask-shape pits in the membrane that resemble the shape of caves (Figure 4.6(c)). They are made of cholesterol-binding protein caveolin with a bilayer full of cholesterol and glycolipids. Besides the clathrin, they are the second most common reported plasma membrane buds. They exist on the surface of many cell types, especially in endothelial cells and adipocytes. CvME is an extremely regulated process that involves multifaceted signaling pathways. The interesting point of CvME is that the particles being internalized could be the main driver of the whole process [44]. The caveolar vesicles from the separation of caveolae from the plasma membrane do not contain any harmful enzymes, so it is noticed that many bacteria and viruses adores this pathway just to avoid enzyme degradation. Thus, it is worthwhile to exploit this mechanism for the research of DNA nanostructures evasion of degradative CME pathway [46]. Furthermore, the caveolar vesicle does not have to go through the endo-lysosomal pathway, it can go to the cavesome. We can say CvME may open a window of opportunity for particles to escape from endosome, which leads to create drug carriers that escape the lysosomal degradation pathway [42]. But in reality, most cells has slower internalization rate through CvME than CME. And the carriage capacity of cavelolar vesicle can only load low fluid-phase volume [44].

4.4.6 *Clathrin- and caveolin-independent endocytosis*

There are also other small structures like caveolae on the surface of plasma membrane, which is call "lipid rafts." Usually these small structures have a diameter of 40–50 nm and feely lay on the cell surface [47]. These "lipid rafts" can be internalized within any endocytic vesicles. This process happens in all type of cells. Unfortunately, the exact mechanisms governing these types of endocytosis are still mysteries; thus we would like to encourage people to exploit this beautiful unknown area.

4.5 DNA Nanostructures in Cellular Environment

The potential of DNA nanostructures to work as drug delivery cargo is related to its structural stability in cellular environment. When we perform the assemblage of DNA nanostructures, we usually do it under room temperature in sterile conditions with appropriate buffer and salt concentrations. However, when the DNA nanostructures goes into blood or even into cells, the environmental conditions are more complicated. *In vivo*, DNA nanostructures will face DNases, higher temperature than room temperature, low salt concentrations (especially Mg^{2+} concentrations are usually lower than 1 mM) and multiple different opsonins and DNA binding proteins. All these complex conditions make it hard for DNA nanostructures to work properly as drug delivery cargo.

4.5.1 *DNA structures in cell lysates*

Cell lysates are mixtures of cellular components that have been homogenized. Lysates does not contain any cell wall, so DNA structures can easily get into a similar environment as inside the cell. According to the research followed by Yan and colleagues that DNA origami can stay in cell lysate and could be extracted and characterized after 12 h of incubation [48]. However, long single/double-stranded nucleic acids cannot be recovered after incubation. Thus, DNA nanostructures are more stable than DNA strands in lysates that constitute a useful setting for exploring DNA nanodevices in physiological conditions.

4.5.2 *DNA structures in serum*

It is also important to study the stability of DNA structures in serum, which is an important path for DNA structures to get to cells. Serum is like lysates; it contains nucleases and lacking stabilizing salts such as magnesium. Conway and colleagues once pointed out that small three-stranded DNA nanostructures which in the shape of triangular prism were more stable than individual strands in serum [49]. In the study, individual strands had a half-life of less than an hour in 10% fetal bovine serum, while the half-life of complete DNA structure was nearly two hours. Furthermore, Perrault and colleagues studied that three different 3D DNA origami in mammalian cell culture media with serum. Their results showed that the stability of DNA origami in serum is related to the design of DNA origami, the presence of Magnesium and the activity of nuclease [50]. They observed that the samples of DNA origami with the addition of 6 mM magnesium did not denatured after one day incubation in that cell culture media, whereas they denatured in the parallel experiment without addition of magnesium. It is interesting that the sample of DNA-origami nanotube is quite stable even without the addition of magnesium in cell culture media. When more than 5% fetal bovine serum was added to the media, all three 3D DNA structures were partially degraded by

DNases; however, when adding actin (a protein that binds to nucleases) to the system, nuclease degradation could be radically reduced. According to other studies, it is shown that tetrahedral DNA nanostructures and DNA origami are more stable than double-stranded DNA in the presence of DNase [51, 52].

All the previous studies suggested that DNA nanostructures are more stable than single/double stranded nucleic acids in physiological environment. Furthermore, some nanostructures can remain their structural integrity in physiological conditions over long time period. It is worthwhile to extend the research in understanding fully the interplay between the functionality and stability of a DNA structure.

4.6 DNA Carriers for Drug Delivery

One of the most important applications of DNA nanostructures is the ability of these fancy structures to work as carriers for drug delivery. In this part of the chapter, we will discuss the possible ways to involve DNA nanostructures in drug delivery system.

The very first design of DNA-origami nanostructures working as molecular containers for drug delivery was single-layer DNA origami. Later on, they were further assembled into 3D nanostructures as shown in Figures 4.7(a) and 4.7(b), a tetrahedron [53] and a hollow cube [54]. After people realize the possibility of DNA structures working as drug delivery carrier, various DNA nanostructures such as nanotubes [55, 56], cages [57], and cubes [58] were designed and synthesized to demonstrate the cellular uptakes of drugs both *in vitro* and *in vivo*. The successful delivery of DNA nanostructures into cells creates new pathway for challenging diverse therapeutic tasks. Jiang *et al.* [59] and Zhang *et al.* [60] demonstrated a way to deliver doxorubicin (DOX) into cells *in vivo* by using rod-like DNA origami or triangular origami as carriers (Figure 4.7(c)). Usually, doxorubicin have low solubility and unfavorable side effects in human body; however, these small drugs can hide behind the GC-rich regions of DNA double helices, which gives great opportunity for DNA nanostructures as new carriers to deliver them. It is worthwhile to mention that; the work of Zhao and colleagues has demonstrated tunable DNA nanostructures for optimal delivery of DOX to human breast cancer cells [61].

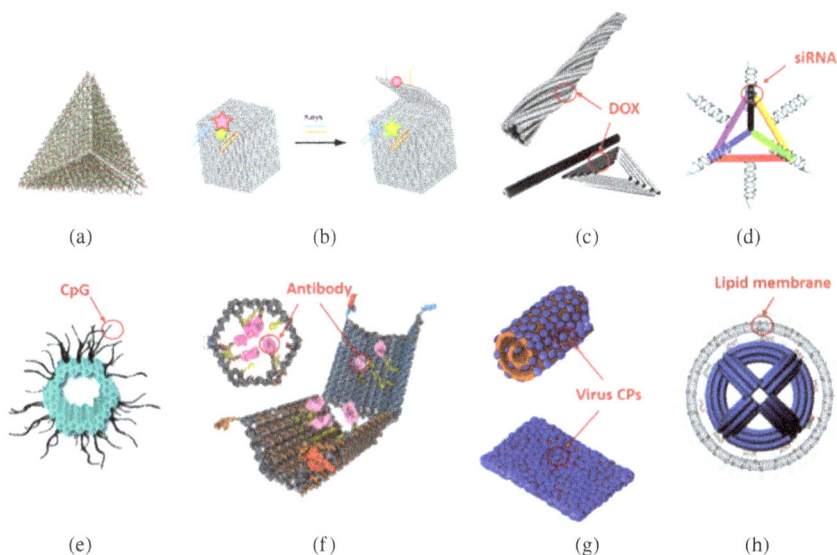

Figure 4.7. Potential DNA-based drug delivery vehicles and devices for triggering cell signaling. (a) a tetrahedron [53] DNA-origami molecular containers. (b) a box with a switchable lid DNA-origami molecular container [54], (c) DNA-origami nanostructures for delivering DOX into cancer cells: an under-twisted rod-like shape [61], a straight rod [59], and a triangle [59, 60]. (d) DNA structures for SiRNA delivery: a cage with cell-targeting ligands (folate or peptide) at the end of the siRNA motifs [63]. (e) DNA structures for CpG-triggered immunostimulation: a DNA-origami tube [65] (f) A smart logic-gated DNA-origami nanorobot to target cells and subsequently display the molecular payload [69]. (g) Rectangular DNA origamis coated with virus CPs for efficient cellular delivery: the complexes can adopt different conformations depending on the added CP concentration [71]. (h) A virus-inspired membrane-encapsulated spherical DNA-origami vehicle for decreasing immune activation and enhancing pharmacokinetic bioavailability [46]. A tetrahedron was adapted with permission from Ref. [53]; Copyright (2009) American Chemical Society. A box with a lid was adapted with permission from Ref. [54]; Copyright (2009) Nature Publishing Group. A straight rod and a triangle were adapted with permission from Ref. [59]; Copyright (2012) American Chemical Society. An under-twisted rod was adapted with permission from Ref. [61]; Copyright (2012) American Chemical Society. siRNA cage was adapted with permission from Ref. [63]; Copyright (2012) Nature Publishing Group. A CpG–DNA-origami tube was adapted with permission from Ref. [65]; Copyright (2011) American Chemical Society. A nanorobot was adapted with permission from Ref. [69]; Copyright (2012) The American Association for the Advancement of Science. Rectangular DNA origamis coated with virus proteins were adapted with permission from Ref. [71]; Copyright (2014) American Chemical Society. A membrane-encapsulated origami was adapted with permission from Ref. [46]; Copyright (2014) American Chemical Society.

In this work, they designed tunable DNA nanostructures that were able to twist and control the release rate of the drug.

DNA nanostructures can be functionalized and then be used for *in vivo* imaging [62]. In addition, DNA nanostructures can control siRNA molecules for silencing genes. For example, Lee and colleagues find that tetrahedral DNA nanostructures modified with folate or peptide can load siRNAs, which can silence target genes in tumors [63] (Figure 4.7(d)). In their research, they noticed that delivery of siRNAs by DNA nanostructures strongly depends on the spatial orientation and the density of the cancer-targeting ligands assembled on the carriers. After few years, Shaw and colleagues demonstrate a way of using DNA-origami "nanocapilars" to adjust the spatial orientation of the membrane-binding ligands, which target to the membrane in cancer cells [64]. Those interesting findings showed the essence of DNA nanostructures of being easily programmed and functionalized is very useful for working as drug delivery carriers.

Afterwards, researchers led their interests in developing DNA-based nanostructures for the application of programmable immunostimulants. Schüller and colleagues successfully conjugate tens of cytosine–phosphate–guanine (CpG) sequences onto tubular DNA origami [65]. (Figure 4.7(e)) They found that using CpG decorated DNA origami induced immunostimulation better than using lipofectamine (a common transfection reagent) delivered CpG. Additionally, different from Lipofectamine, CpG-origami are non-cytotoxic complexes. In 2015, Sellner and colleagues introduced successful experiments with CpG-coated origamis *in vivo* [56]. Moreover, several researches about different DNA nanostructures as carriers for immunostimulatory CpG motifs were successfully conducted, such as tetrahedral DNA cages [66], polypod-like structured DNA [67] and DNA dendrimers [68]. All those successful results give us a great hope of using DNA nanostructures as new drug carriers in cancer therapy. One of the most remarkable researches needs to be point here is the logic-gate DNA-origami nanorobot developed by Douglas *et al.* [69]. The nanorobot is a highly sophisticated delivery system that can sense the surface proteins of the target cells and then trigger programmed aptamer-based logic-gates thus selectively transport the molecular payload to the cells (Figure 4.7(f)).

Although all those great achievements in this field, we still need to find an efficient transfection method for DNA nanostructure drug carriers. Due to the reason that DNA origami are polar structures due to the backbone of nucleic acid, they open need modifications of peptide, cationic polymer or lipid to successfully transport into cells. However, Brglez and colleagues showed another possible way to improve the delivery efficiency of DNA nanostructures. The surface properties of DNA origami can be modified by the specific DNA intercalators thus subsequently improve the transfection efficiency [70]. Furthermore, Mikkilä and colleagues decorated a single rectangular DNA-origami sheet with virus capsid proteins (CPs) from cowpea chlorotic mottle virus (CCMV) [71]. The CP-origami complexes showed different morphologies, a roll or a flat plate, which depends on the amount of CPs on DNA-origami (Figure 4.7(g)). It is noticed that the transfection rate increased significantly compares to the bare DNA-origami structure of the experiments on human cell line HEK293. Importantly, they didn't observe any toxicity in their experiments. Inspired by the viruses, Perrault and Shih [46] creatively encapsulate spherical DNA-origami structures in conjugated (PEGylated) lipid membranes (Figure 4.7(h)). This invention smartly protects DNA-origami nanostructures from nuclease digestion, and moreover, the immune activation *in vivo* experiments was significantly decreased.

4.7 Conclusion and Perspectives

Given the fundamental importance of DNA in biology, it is expected that DNA nanotechnology could represent a real breakthrough in life sciences, enabling a plethora of unprecedented, complex and exciting applications. In this chapter, we showed that the extremely high flexibility in the design, high cell permeability, biocompatibility and spatial positioning of self-assembled DNA nanostructures permit them to be the ideal systems as carriers for molecular payloads delivery, such as drugs, antibodies and siRNA. We showed of ligands to be a powerful technique to mildly decorate 2D or also 3D surfaces with capture DNA aiming at immobilizing multiple proteins on the same surface or manipulate single cells by triggering extracellular or intracellular responses. Future developments of this approach, thanks to the integration of top-down nanofabrication approaches, will permit to immobilize gradients of ligands or mimicking the topography and functionality of the extracellular matrix.

On the other hand, the field of DNA-origami has shown significant development in the last decade and increasing numbers of labs and researchers are now exploring such technology, with many potential exciting applications being described. In particular, applications in drug delivery by DNA nanostructures can be implemented in practice by understanding of the pharmacokinetics and biodistribution immunostimulatory properties of DNA nanostructures and finally by decreasing the production costs, especially for large nanostructures such as DNA-origami. Researches in this field will have to overcome the challenges associated with the translation of these technologies to the clinic and widespread biotechnology testing.

References

1. Chen, N., He, Y., Su, Y., Li, X., Huang, Q., Wang, H., Zhang, X., Tai, R., and Fan, C. (2012). The cytotoxicity of cadmium-based quantum dots. *Biomaterials*. **33**, 1238–11244.
2. Magrez, A., Kasas, S., Salicio, V., Pasquier, N., Seo, J.W., Celio, M., Catsicas, S.B. Schwaller, B., and Forró, L. (2006). Cellular toxicity of carbon-based nanomaterials. *Nano. Lett*. **6**, 1121–1125.
3. Lv, H., Zhang, S., Wang, B., Cui, S., and Yan, J. (2006). Toxicity of cationic lipids and cationic polymers in gene delivery. *J. Control. Release*. **114**, 100–109.
4. Petros, R.A., and DeSimone, J.M. (2010). Strategies in the design of nanoparticles for therapeutic applications. *Nat. Rev. Drug Discov*. **9**, 615–627.
5. Choi, H.S., Liu, W., Liu, F., Nasr, K., Misra, P., Bawendi, M.G., and Frangioni, J.V. (2010). Design considerations for tumour-targeted nanoparticles. *Nat. Nanotechnol*. **5**, 42–47.
6. Davis, M.E., Chen, Z., and Shin, D.M. (2008). Nanoparticle therapeutics: An emerging treatment modality for cancer. *Nat. Rev. Drug Discovery*. **7**, 771–782.
7. Lee, H., Lytton-Jean, A.K., Chen, Y., Love, K.T., Park, A.I., Karagiannis, E.D., Sehgal, A., Querbes, W., Zurenko, C.S., and Jayaraman, M. (2012). Molecularly self-assembled nucleic acid nanoparticles for targeted *in vivo* siRNA delivery. *Nat. Nanotechnol*. **7**, 389–393.

8. Lyu, Y., Chen, G., Shangguan, D., Zhang, L., Wan, S., Wu, Y., Zhang, H., Duan, L., Liu, C., You, M., Wang, J., and Tan, W. (2016). Generating cell targeting aptamers for nanotheranostics using cell-SELEX. *Theranostics.* **6**, 1440–1452.

9. Bing, T., Yang, X., Mei, H., Cao, Z., and Shangguan, D. (2010). Conservative secondary structure motif of streptavidin-binding aptamers generated by different laboratories. *Bioorg. Med. Chem.* **18**, 1798–1805.

10. Wang, C., Zhang, M., Yang, G., Zhang, D., Ding, H., Wang, H., *et al.* (2003). Single-stranded DNA aptamers that bind differentiated but not parental cells: subtractive systematic evolution of ligands by exponential enrichment. *J. Biotechnol.* **102**, 15–22.

11. Shangguan, D., *et al.* (2006). *In vitro* selection with artificial expanded genetic information systems. *Proc. Natl. Acad. Sci. USA.* **103**, 11838–11843.

12. Jiang G., *et al.* (2012). PTK7: A new biomarker for immunophenotypic characterization of maturing T cells and T cell acute lymphoblastic leukemia. *Leuk. Res.* **36**, 1347–1353.

13. Meng, H.-M., Liu, H., Kuai, H., Peng, R., Mo, L., and Zhang, X.B. (2016). Aptamer-integrated DNA nanostructures for biosensing, bioimaging and cancer therapy. *Chem. Soc. Rev.* **45**, 2583–2602.

14. Pei, H., Liang, L., Yao, G., Li, J., Huang, Q., and Fan, C. (2012). Reconfigurable three-dimensional DNA nanostructures for the construction of intracellular logic sensors. *Angew. Chem. Int. Ed. Engl.* **51**, 9020–9024.

15. Wu, C., Cansiz, S., Zhang, L., Teng, I.T., Qiu, L., Li, J., Liu, Y., Zhou, C., Hu, R., Zhang, T., Cui, C., Cui L., and Tan, W. (2015). A nonenzymatic hairpin DNA cascade reaction provides high signal gain of mrna imaging inside live cells. *J. Am. Chem. Soc.* **137**, 4900–4903.

16. Pantel, K., and Speicher, M.R. (2016). The biology of circulating tumor cells. *Oncogene,* **35**, 1216–1224.

17. Qu, L., Xu, J., Tan, X., Liu, Z., Xu, L., and Peng, R. (2014). Dual-aptamer modification generates a unique interface for highly sensitive and specific electrochemical detection of tumor cells. *ACS Appl. Mater. Interfaces,* **6**, 7309–7315.

18. Li, S., Chen, N., Zhang Z., and Wang, Y. (2013). Endonuclease-responsive aptamer-functionalized hydrogel coating for sequential catch and release of cancer cells. *Biomaterials.* **34**, 460–469.

19. Ma, H., Liu, J., Ali, M.M., Mahmood, M.A., Labanieh, L., Lu, M., Iqbal, S.M., Zhang, Q., Zhao W., and Wan, Y. (2015). Nucleic acid aptamers in cancer research, diagnosis and therapy. *Chem. Soc. Rev.* **44**, 1240–1256.

20. Wang, R., Zhu, G., Mei, L., Xie, Y., Ma, H., Ye, M., Qing F.L., and Tan, W. (2014). Automated modular synthesis of aptamer-drug conjugates for targeted drug delivery. *J. Am. Chem. Soc.* **136**, 2731–2734.

21. Zhu, G., Zheng, J., Song, E., Donovan, M., Zhang, K., Lin, C., and Tan, W. (2013). Self-assembled, aptamer-tethered DNA nanotrains for targeted transport of molecular drugs in cancer theranostics. *Proc. Natl. Acad. Sci. USA.* **110**, 7998–8003.

22. McNamara, J.O., Andrechek, E.R., Wang, Y., Viles, K.D., Rempel, R.E., Gilboa, E., Sullenger, B.A., and Giangrande, P.H. (2006). Cell type-specific delivery of siRNAs with aptamer-siRNA chimeras. *Nat. Biotechnol.* **24**, 1005–1015.

23. Lv, Y., Peng, R., Zhou, Y., Zhang, X.B., and Tan, W. (2016). Catalytic self-assembly of a DNA dendritic complex for efficient gene silencing. *Chem. Commun.* **52**, 1413–1415.

24. Krieg, A.M. (2002). CpG motifs in bacterial DNA and their immune effects. *Ann. Rev. Immunol.* **20**, 709–760.

25. Liu, X., Xu, Y., Yu, T., Clifford, C., Liu, Y., Yan, H., and Chang, Y. (2012). A DNA nanostructure platform for directed assembly of synthetic vaccines. *Nano. Lett.* **12**, 4254–4259.

26. Xiong, X., Liu, H., Zhao, Z., Altman, M.B., Lopez-Colon, D., Yang, C.J., Chang, L.J., Liu, C., and Tan, W. (2013). DNA aptamer-mediated cell targeting. *Angew. Chem., Int. Ed.* **52**, 1472–1476.

27. Arrabito, G., and Pignataro, B. (2012). Solution processed micro- and nano-bioarrays for multiplexed biosensing. *Anal. Chem.* **84**, 5450–5462.

28. Meyer, R., *et al.* (2014). Advances in DNA-directed Immobilization. *Curr. Opin. Chem. Biol.* **18**, 8–15.

29. Niemeyer, C.M. (2010). Semisynthetic DNA–protein conjugates for biosensing and nanofabrication. *Angew. Chem. Int. Ed.* **49**, 1200–1216.

30. Chandra, R.A., Douglas, E.S., Mathies, R.A., Bertozzi, C.R., Francis, M.B. (2006). Programmable cell adhesion encoded by DNA hybridization. *Angew. Chem. Int. Ed. Engl.* **45**, 896–901.

31. Reisewitz, S., Schroeder, H., Tort, N., Edwards, K.A., Baeumner, A.J., and Niemeyer, C.M. (2010). Capture and culturing of living cells on microstructured DNA substrates. *Small.* **6**, 2162–2168.

32. Arrabito, G., *et al.* (2013). Biochips for cell biology by combined dip-pen nanolithography and DNA-Directed protein immobilization. *Small.* **9**, 4243–4249.

33. Gandor, S., *et al.* (2013). A Protein-Interaction Array Inside a Living Cell. *Angew. Chem. Int. Ed. Engl.* **52**, 4790–4794.

34. Arrabito, G., *et al.* (2014). Configurable low-cost plotter device for fabrication of multi-color sub-cellular scale microarrays. *Small.* **10**, 2870–2876.
35. Huo, F., Zheng, Z., Zheng, G., Giam, L.R., Zhang, H., and Mirkin C.A. (2008). Polymer pen lithography. *Science.* **321**, 1658–1660.
36. Angelin, A., Weigel, S., Garrecht, R., Meyer, R., Bauer, J., Kumar, R.K., Hirtz M., and Niemeyer, C. (2015). Multiscale origami structures as interface for cells. *Angew. Chem. Int. Ed. Engl.* **54**, 15813–15817.
37. Gratton, S.E., Ropp, P.A., Pohlhaus, P.D., Luft, J.C., Madden, V.J., *et al.* (2008). The effect of particle design on cellular internalization pathways. *Proc. Natl. Acad. Sci. USA.* **105**, 11613–11618.
38. Anderem, A., and Underhill, D.M. (1999). Mechanism of phagocytosis in macrophages. *Annu. Rev. Immunol.* **17**, 593–623.
39. Owens, D.E., III, and Peppas, N.A. (2006). Opsonization, biodistribution, and pharmacokinetics of polymeric nanoparticles. *Int. J. Pharm.* **307**, 93–102.
40. Korn, E.D., and Weisman, R.A. (1967). Phagocytosis of latex beads by acanthamoeba. *J. Cell Biol.* **34**, 219–227.
41. Vonarbourg, A., Passirani, C., Saulnier, P., Simard, P., Leroux, J.C., and Benoit, J.P. (2006). Evaluation of pegylated lipid nanocapsules versus complement system activation and macrophage uptake. *J. Biomed. Mater. Res. Part A.* **78**, 620–628.
42. Hillaireau, H., and Couvreur, P. (2009). Nanocarriers' entry into the cell: Relevance to drug delivery. *Cell. Mol. Life Sci.* **66**, 2873–2896.
43. Mukherjee, S., Ghosh, R.N., and Maxfield, F.R. (1997). Endocytosis. *Physiol. Rev.* **77**, 759–803.
44. Conner, S.D., and Schmid, S.L. (2003). Regulated portals of entry into the cell. *Nature.* **422**, 37–44.
45. Strømhaug, P., Berg, T., Gjøen, T., and Seglen, P. (1997). Differences between fluid-phase endocytosis (pinocytosis) and receptor-mediated endocytosis in isolated rat hepatocytes. *Eur. J. Cell Biol.* **73**, 28–39.
46. Perrault, S.D., and Shih, W.M. (2014). Virus-inspired membrane encapsulation of dna nanostructures to achieve *in vivo* stability. *ACS Nano.* **8**, 5132–5140.
47. Edidin, M. (2001). Shrinking patches and slippery rafts: Scales of domains in the plasma membrane. *Trends Cell Biol.* **11**, 492–496.
48. Mei, Q., *et al.* (2011). Stability of DNA origami nanoarrays in cell lysate. *Nano. Lett.* **11**, 1477–1482.
49. Conway, J.W., McLaughlin, C.K., Castor, K.J., and Sleiman, H. (2013). DNA nanostructure serum stability: Greater than the sum of its parts. *Chem. Commun.* **49**, 1172–1174.

50. Hahn, J., Wickham, S.F.J., Shih, W.M., and Perrault, S.D. (2014). Addressing the instability of DNA nanostructures in tissue culture. *ACS Nano.* **8**, 8765–8775.
51. Keum, J.W., and Bermudez, H. (2009). Enhanced resistance of DNA nanostructures to enzymatic digestion. *Chem. Commun.* **45**, 7036–7038.
52. Castro, C.E., *et al.* (2011). A primer to scaffolded DNA origami. *Nat. Meth.* **8**, 221–229.
53. Ke, Y., Sharma, J., Liu, M., Jahn, K., Liu, Y., and Yan, H. (2009). Scaffolded DNA origami of a DNA tetrahedron molecular container. *Nano. Lett.* **9**, 2445–2447.
54. Andersen, E.S., *et al.* (2009). Self-assembly of a nanoscale DNA box with a controllable lid. *Nature.* **459**, 73–76.
55. Ko, S.H., Liu, H., Chen, Y., and Mao, C. (2008). DNA nanotubes as combinatorial vehicles for cellular delivery. *Biomacromolecules.* **9**, 3039–3043.
56. Sellner, S., Kocabey, S., Nekolla, K., Krombach, F., Liedl, T., and Rehberg, M. (2015). DNA nanotubes as intracellular delivery vehicles *in vivo*. *Biomaterials.* **53**, 453–463.
57. Walsh, A.S., Yin, H., Erben, C.M., Wood, M.J., and Turberfield, A.J. (2011). DNA cage delivery to mammalian cells. *ACS Nano.* **5**, 5427–5432.
58. Bujold, K.E., *et al.* (2014). Sequence-responsive unzipping DNA cubes with tunable cellular uptake profiles. *Chem. Sci.* **5**, 2449– 2455.
59. Jiang, Q., *et al.* (2012). DNA origami as a carrier for circumvention of drug resistance. *J. Am. Chem. Soc.* **134**, 13396–13403.
60. Zhang, Q., Jiang, Q., Li, N., Dai, L., Liu, Q., Song, L., Wang, J., Li, Y., Tian, J., Ding, B., and Du, Y. (2014). DNA origami as an *in vivo* drug delivery vehicle for cancer therapy. *ACS Nano.* **8**, 6633–6643.
61. Zhao, Y.X., Shaw, A., Zeng, X., Benson, E., Nystrom, A.M., and Hogberg, B. (2012). DNA origami delivery system for cancer therapy with tunable release properties. *ACS Nano.* **6**, 8684–8691.
62. Bhatia, D., Surana, S., Chakraborty, S., Koushika, S.P., and Krishnan, Y. (2011). A synthetic icosahedral DNA-based host–cargo complex for functional *in vivo* imaging. *Nat. Commun.* **2**, 339.
63. Lee, H., *et al.* (2012). Molecularly self-assembled nucleic acid nano- particles for targeted *in vivo* siRNA delivery. *Nat. Nanotech.* **7**, 389–393.
64. Shaw, A., *et al.* (2014). Spatial control of membrane receptor function using ligand nanocalipers. *Nat. Methods.* **11**, 841–846.
65. Schüller, V.J., Heidegger, S., Sandholzer, N., Nickels, P.C., Suhartha, N.A., Endres, S., Bourquin, C., and Liedl, T. (2011). Cellular immunostimulation

by CpG-sequence-coated DNA origami structures. *ACS Nano.* **5**, 9696– 9702.

66. Li, J., Pei, H., Zhu, B., Liang, L., Wei, M., He, Y., Chen, N., Li, D., Huang, Q., and Fan, C. (2011). Self-assembled multivalent DNA nanostructures for noninvasive intracellular delivery of immunostimulatory CpG oligonucleotides. *ACS Nano.* **5**, 8783–8789.

67. Mohri, K., *et al.* (2012). Design and development of nanosized DNA assemblies in polypod-like structures as efficient vehicles for immunostimulatory CpG motifs to immune cells. *ACS Nano.* **6**, 5931–5940.

68. Mohri, K., *et al.* (2015). Self-assembling DNA dendrimer for effective delivery of immunostimulatory CpG DNA to immune cells. *Biomacromolecules.* **16**, 1095–1101.

69. Douglas, S.M., Bachelet, I., and Church, G.M. (2012). A logic-gated nanorobot for targeted transport of molecular payloads. *Science.* **335**, 831–834.

70. Brglez, J., Nikolov, P., Angelin, A., and Niemeyer, C.M. (2015). Designed intercalators for modification of DNA origami surface properties. *Chem. Eur. J.* **21**, 9440–9446.

71. Mikkilä, J., Eskelinen, A.P., Niemelä, E.H., Linko, V., Frilander, M.J., Törmä, P., and Kostiainen, M.A. (2014). Virus-encapsulated DNA origami nano- structures for cellular delivery. *Nano. Lett.* **14**, 2196–2200.

Chapter 5

Targeting G-quadruplex DNA as Potential Anti-cancer Therapy

Riccardo Bonsignore*,‡, Elisa Trippodo† and Giampaolo Barone†

*School of Chemistry, Cardiff University
Park Place, Cardiff, UK
†Department of Biological, Chemical and Pharmaceutical
Sciences and Technologies University of Palermo
Viale delle Scienze, Ed. 17, 90128 Palermo, Italy
‡BonsignoreR@cardiff.ac.uk

This chapter provides an introduction and also a review on non-canonical DNA structures such as guanine-quadruplexes (G-quadruplexes) and their targeting by small molecules for applications in Pharmacology. The articles considered in this chapter are mainly from 2016.

5.1 Introduction

When referring to the biological role of DNA in living beings it is easy to relate it to the hard drive of a personal computer: as the latter, the polynucleotide stores all the information needed for a proper cellular life. Any molecule capable to interfere with the DNA, blocking its function, may therefore become lethal for a cell. For such a reason the polynucleotide has been widely considered an important target to halt cell proliferation, especially in life-threating diseases as cancer. Nonetheless, DNA-binders, being unable to discriminate between the DNA of healthy and cancerous cells, have been always characterized by several side effects. This, together with the recent finding of more specific targets, e.g. proteins, shifted the interests of the researchers towards the latter.

Despite this, recent discoveries about the involvement of non-canonical DNA secondary structures in several pathologies — cancer, above all — have again turned on the lights on the polynucleotide as a potential target in anti-cancer therapies. The hypothesis is that, when a molecule is capable to interact with a specific secondary structure of the polynucleotide, different from the canonical double helix, it could affect mainly cancerous cells causing fewer side effects. Among these structures, G-quadruplexes sequences have been widely found to be involved in cancer progression and therefore are nowadays considered as potential targets of small binders.

In the present chapter we intended to focus our attentions on the advances in G-quadruplex DNA targeting by small molecules in the year 2016, mainly in an anti-cancer view. For such a reason, the first two paragraphs will give basic information about the structural and biological aspects of G-quadruplex DNA. These themes have been widely and properly discussed in several reviews and books, which we address the readers to refer to for wider and more detailed information [1–3]. In the second part of the chapter we will introduce the reader to recent discoveries of G-quadruplex DNA binders, discussing typical design processes.

5.2 Structural Details

The propensity of guanine-enriched nucleic acids to self-associate has been recognized since 1910 when Bang showed these sequences could form gel-like substances in water [4]. This evidence was later supported by spectroscopic studies that allowed Ralph *et al.* [5], to realize that the optical density of tri- and tetradeoxyguanosines changed with temperature. In particular a thermal transition point was evident, pointing out the existence of an ordered secondary structure. Nonetheless, the real turning point in the understanding the nature of G-quadruplex DNA was found still in the 1962 when Gellert *et al.*, through X-ray diffraction studies on gels of 3'- and 5'-guanosine monophosphate, demonstrated that their arrangement was similar to that of double-helical DNA [6]. Notably, according to the authors, the high stability of this secondary nucleotide structure could be only explained by the hydrogen-bonding of four guanine bases.

The building block of a G-quadruplex DNA is indeed made up of a tetrad of guanines linked by Hoogsteen hydrogen bonds involving the N1, N2, N7 and O6 atoms of each guanine base (Figure 5.1(a)). The final quadruplex structure is stabilized by $\pi-\pi$ interactions among the adjacent stacked tetrads (Figure 5.1(b)), whose misalignment (Figure 5.1(c)) gives rise to a quadruple helical-like structure. Notably, the formation of these square planar platforms is driven by the presence of a monovalent cation, typically K^+ or Na^+, equidistant from the eight O6 atoms of two adjacent tetrads [7, 8], and forming a ionic channel in the center of the whole G-quadruplex (Figure 5.1(c)).

As double-helical DNA, also G-quadruplex DNA presents a high polymorphic nature: as a matter of fact, these structures can easily adopt different conformations depending on the solution conditions and on the nature of monovalent ion. In this respect, due to the higher intracellular concentration of K^+ than Na^+ (about 140 mM vs. 10 mM), the potassium driven structure is considered to be biologically more relevant [8]. Moreover, each of the guanine residues can adopt either a *syn-* or an *anti-*glycosidic conformation, giving rise to 16 possible conformations *per* tetrad and many more when considering a whole G-quadruplex [9].

G-quadruplex DNA can arise from the folding of a single (intramolecular) guanine enriched G-tract, or from two up to four guanine-enriched separate strands, forming an intermolecular G-quadruplex [3]. Moreover, according to the orientations of the strands, G-quadruplexes can be classified as parallel, antiparallel and mixed, or hybrid, with several possible connecting tracts (loops), thus increasing their polymorphism.

From the considerations above it becomes clear that the knowledge of the *in vivo* structure of the G-quadruplex DNA is an important challenge in order to design optimal molecules with enhanced biological properties, capable to bind these structure for therapeutic purposes, as it will be

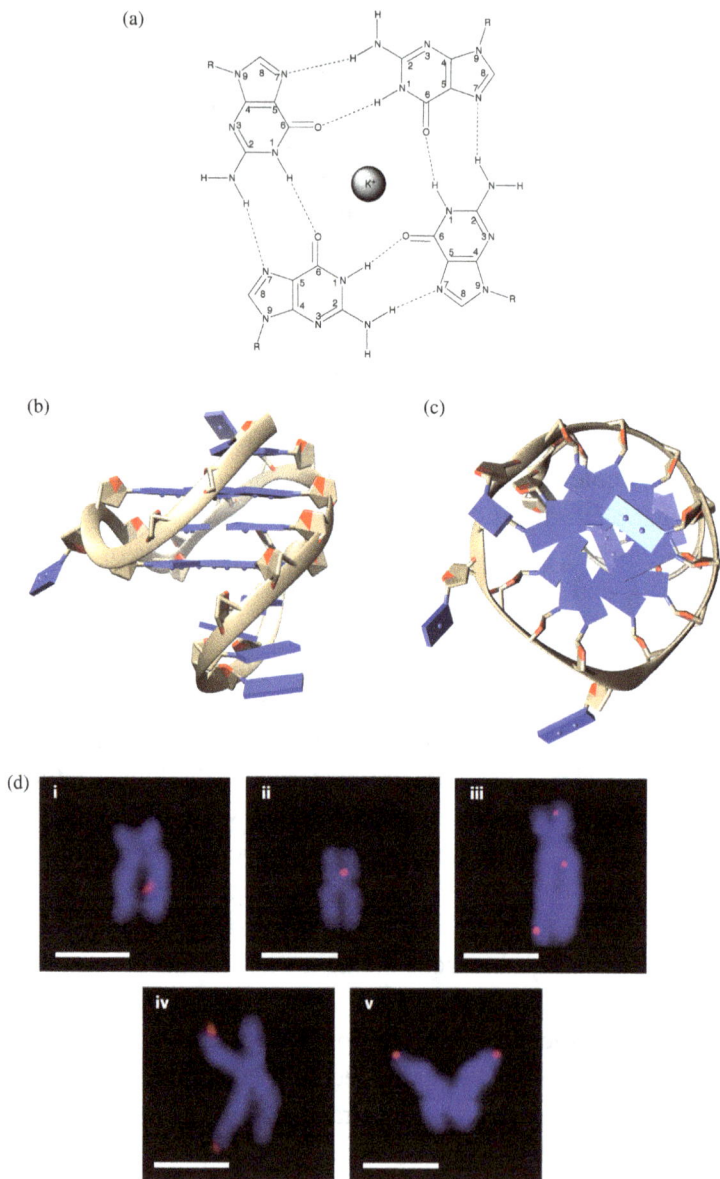

Figure 5.1. (a) Representation of a G-tetrad with Hoogsteen hydrogen bonds; (b) front and (c) top view of a G-quadruplex from c-KIT DNA (Adapted with permission from A.T. Phan *et al. J. Am. Chem. Soc.*, 2007, 129, 4386–4392. Copyright (2007) American Chemical Society); (d) *in vivo* visualization of G-quadruplex DNA (red dots) by means of a fluorescent antibody. Adapted with permission from Ref. [11]. Copyright (2013) Nature Publishing Group.

discussed in the following paragraphs. In such a context, X-ray crystallography and NMR spectroscopy have been widely and predominantly used to characterize structural aspects of G-quadruplex DNA.

5.3 Biological Involvement

More than 30 years ago, Aaron Klug stated that since G-quadruplexes could form so easily *in vitro*, Nature should have found a way to use them in Biology. This statement postulated the involvement of these structures in complex biological functions, despite their observation *in vivo* was not yet achieved. The development of specific antibodies directed against G-quadruplex folded DNA represented the real turning point to confirm their existence and to localize them in living cells. For example, by *in situ* immuno-staining, Schaffitzel and coworkers in the 2001 were able to localize and visualize the telomeric G-quadruplex DNA in the micronuclei of the ciliate *Stylonychia lemnae* [10]. As for human DNA, the direct *in vivo* visualization of the G-quadruplex DNA was recently achieved by Balasubramanian's group in 2013 by means of a specific engineered antibody [11]. As shown in Figure 5.1(d), stable G-quadruplex motifs were found in telomeres, gene bodies and gene regulatory regions [12]. Interestingly, bioinformatic analysis of the human genome previously revealed that over 300,000 human DNA sequences might fold into a G-quadruplex [13]: these sequences are not-randomly distributed but mainly located in telomeric regions and in proximity of gene promoters. In particular, the development of next-generation sequencing and polymerase stop assays have allowed the identification of more than 700,000 human G-quadruplexes, of which almost 450,000 have not been predicted before, such as BRCA1 and BRCA2 genes [2]. Overall, these findings point out the regulatory role played by G-quadruplex DNA, in the transcription of genes and in the replication of human genome, as it will be discussed in the next paragraphs.

5.3.1 *G-quadruplexes in human oncogene promoters*

The first evidences of "altered DNA conformations" in the nuclease hypersensitivity of promoter elements were found in 1982, when Larsen *et al.* investigated the chicken β-globulin gene [14]. Later, several G-enriched promoter sequences, potentially able to fold into G-quadruplexes, were identified in a wide amount of human genes involved in growth and pro-liferation processes, including many of the human oncogenes [15]. For example, the occurrence of these structures has been found in genes such as human Vascular Endothelial Growth Factor (VEGF) [16], c-MYC [17, 18], Hypoxia-Inducible Factor-1α (HIF-1α) [19], B-Cell Lymphoma-2 (BCL-2) [20–22], sMtCK [23], human and mouse Kirsten RAt Sarcoma (KRAS) [24], human c-KIT [25, 26], REarranged during Transfection (RET) [27] and Platelet-Derived Growth Factor-A (PDGF-A) [28]. Among all of them, the c-MYC oncogene was the first human gene systematically studied as a G-quadruplex target for anti-cancer ther-apy, due to its importance in cancer proliferation and also because its protein, MYC, is quite unstable and therefore hard to be targeted by any drug [1].

These discoveries raised the interest of several scientists, since the stabilization of a G-quadruplex motif in promoter oncogenes could allow the down regulation of genes playing key roles in cancer progression. The assumption is that, when a small molecule binds to a G-quadruplex folded DNA sequence, RNA polymerase and transcription factors, as well, are not anymore able to recognize these DNA sequences: as a consequence a mRNA will not be produced and the amount of the related-synthesized proteins will decrease. Furthermore, this effect would be ideal when the binding ligand is capable of selectively recognizing the targeted G-quadruplex of the overexpressed gene sequence, rather than any other G-quadruplex or duplex DNA. Nonetheless, such specificity still remains one of the major challenges and main goals of the "G-quadruplex target-ing" research.

5.3.2 *G-quadruplexes in telomeres*

Human telomeres, at the terminus of chromosomes, are characterized by single-strand overhangs of repeated TTAGGG sequences. Telomeric DNA physiologically shortens during each replication cycle as a consequence of the inability of the DNA-replication machinery of the cell to fully repli-cate these ends [29]. Therefore, in the absence of compensating mecha-nisms, telomeres in normal cells progressively shorten until they reach the limit point — called Hayflick limit — when cells respond by ceasing replication and entering senesce, normally leading to apoptosis [30]. Nevertheless, some cells, such as stem and germ cells, have the need to elongate the telomeres ends, thus keeping their unlimited proliferative capabilities. In particular, for this purpose, these cells synthesize an enzyme named telomerase, owning reverse transcriptase activity. This enzyme is not expressed in other healthy cells, causing a loss of 50–200 bases at the telomeric region *per* replication cycle. Interestingly, telomerase has been also found to be expressed in most of human cancers (85–90%) [31] granting them unlimited proliferation ability and consequent immortalization. On the other hand, the occurrence of G-quadruplex secondary structures in these G-rich tracts allows designing specific G-quadruplex binders and stabilizers: once stabilized, indeed, telomerase should not be anymore able to recognize its natural substrate and therefore no elongation of the telomeres would take place. As a consequence, cancer cells might enter senescence and probably undergo apoptosis as a healthy human cell. However, the employment of telomerase inhibitors is limited by the presence of alternative lengthening mechanisms (ALT) in cancer cells, which maintain the integrity of telomeres and, consequently, their unlimited replicative ability [32]. Notably, recent findings seem to point out that the interaction between stabilizing molecules and telomeric G-quadruplexes might bring to the uncapping of telomeres from constitutive proteins, damaging these ends. In such a context it seems that the inhibition of the telomerase activity might play a secondary but still precious role [2].

It is worth noticing that the same specificity-issues reported in Section 5.3.1 are also related to the design and synthesis of telomeric G-quadruplex binders, thus pushing scientists to exploit structure-activity relationships to achieve a specific interaction with the aimed G-quadruplex, goal yet still far to be reached.

5.4 Recent Advances in Targeting G-quadruplex DNA

A multidisciplinary approach is nowadays used to assess the DNA-binding properties of G-quadruplex stabilizers: among the experimental techniques exploited, those such as UV–Vis, Circular Dichroism (CD), NMR spectroscopies, as well as ESI–MS and FRET melting assays, are

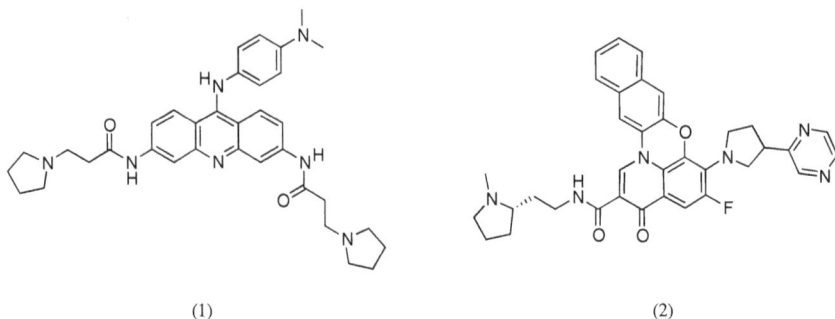

(1)   (2)

Figure 5.2.   Molecular structures of BRACO-19 (**1**) and Quarfloxin (**2**).

commonly used to collect clues of the G-quadruplex binding capabilities of newly synthesized compounds. These results are often corroborated by computational investigations, through molecular docking and/or molecular dynamics (MD) simulations, as well as by biological studies *in vitro*, such as cell cycle analysis, Western Blotting of the expression of the gene, telomerase activity and antiproliferative assays.

Noteworthy, among the newly synthesized molecules with G-quadruplex stabilizing abilities, the discoveries of compounds such as BRACO-19 (**1**) [33] and Quarfloxin (**2**) [34] represent a milestone to guide the design of new molecules (see Figure 5.2).

Since then, indeed, researchers have been exploring the suitable features that a compound must possess in order to preferentially bind G-quadruplex rather than double-helical DNA. One of the widest explored binding modes is the **top-stacking** of a molecule on one of the two external faces (3′ or 5′) of the G-quadruplex motifs. In such a context, two main structural features are needed for a molecule to establish this kind of interaction:

(i) a flat planar π-system, allowing π–π interactions between the binder and the top quartet of the G-quadruplex;

(ii) the presence of polar side chains, possibly positively charged at physiologic pH, able to interact with the negatively charged sugar–phosphate backbone in the DNA grooves thus reinforcing the interaction with the G-quadruplex.

On the other hand, some compounds have been designed as **G-quadruplex groove binders**. The latter feature has been exploited to induce a certain degree of selectivity among G-quadruplex sequences, and to increase the affinity for particular structural properties possessed by certain G-quadruplex conformations. For example, the NMR structure of G-quadruplex DNA from c-KIT sequence has shown to possess a deep hydrophobic cleft never found in other G-quadruplexes or in duplex DNA. Therefore, designing a molecule capable to structurally fit with such groove might reveal to be a powerful strategy to achieve a binding specificity that could reduce side effects in anti-cancer therapies.

For the same reason, many research groups have reported the increase in G-quadruplex binding capabilities of molecules in which a **metal ion** is complexed by a suitable organic ligand. In fact, the presence of a cationic metal ion usually increases water solubility, which is necessary since the final interaction occurs in cell environments, where water is the solvent. Moreover, the presence of a metal ion allows a fine-tuning of the structural properties of the complex, which can be exploited in order to increase the interaction with the grooves or with the guanine tetrads. In the latter case, it is presumable that the presence of a metal ion drives the positioning on the top face of the G-quadruplex, by aligning itself on top of the internal K^+ or Na^+ channel.

For such a reason the following chapter is dedicated to the literature findings of the year 2016, concerning organic ligands and metal-complexes as G-quadruplex binders.

5.4.1 *Targeting G-quadruplex DNA with organic binders*

Small isoquinoline alkaloids, found in several plants, have been widely used by traditional medicine for the treatment of several diseases, from fever to cancer [35, 36]. Among them, derivative **4** has been found to preferentially bind human G-quadruplex DNA [37–40]. In detail, Malhotra and coworkers, after setting up a convenient route to synthesize dihydrochelerythrine (**3**), were able to assess its G-quadruplex binding capabilities after a comparison with the parental alkaloid **4**, using biophysical and biochemical approaches [41]. In particular, ligand induced thermal stabilization experiments through CD spectroscopy (CD melting experiments) were carried out in order to unravel the stabilizing capabilities of **3** and **4** towards c-MYC and c-KIT1 G-quadruplexes. When the DNA secondary structure is stabilized by means of a binder it is in fact expected that the amount of the energy needed to separate its filaments increases with respect to the polynucleotide alone: thus an increase of the melting temperature of the DNA occurs. Malhotra *et al.* found that, on c-MYC G-quadruplex, **3** caused a larger stabilization than **4**, and that this order was reversed for c-KIT1 G-quadruplex. Nonetheless, no interaction with duplex DNA was found, since its melting temperature remained constant upon addition of both derivatives. Computational investigations on the binding mode of **3** to the c-MYC G-quadruplex revealed π-stacking to both the top both the bottom quartets of the G-quadruplex as the main binding mode. Lastly, biological assays showed that **3** exhibited a 30% inhibitory effect on the proliferation of the only A431 cancer cell line. This result pointed out interesting cancer-tissue specificity for this derivative [41].

Similar c-MYC selectivity, over duplex DNA and other G-quadruplexes, was found by Kumar and coworkers in the early 2016. The authors, indeed, employing one-pot modular click reactions were capable to synthesize two dansyl-guanosine conjugates, with the derivative **5** showing the most interesting G-quadruplex binding properties [42] (see Figure 5.3). NMR spectroscopic titrations of c-MYC G-quadruplex in the presence of increasing amounts of **5** revealed line broadening and chemical-shift perturbations of the NMR spectra of the G-quadruplex alone. This result allowed the authors to suggest that **5** preferentially binds to 3′-end-capping structure and to the groove region of the G-quadruplex. Moreover, antiproliferative assays were carried out on HT1080 cell lines, which are characterized by an overexpression of c-MYC oncogene. These studies showed that compound **5** could cause a dose-dependent inhibition in cell proliferation, thus addressing its *in vitro* anticancer properties. Noteworthy, by means of other biological approaches — such as quantitative real-time PCR (qRT-PCR) — Kumar *et al.* were able to demonstrate a gradual decrease, in the presence of increasing concentration of **5**, of the amount of c-MYC transcript. Despite the exact mechanism underlying this repression has not yet been found, this result can be considered

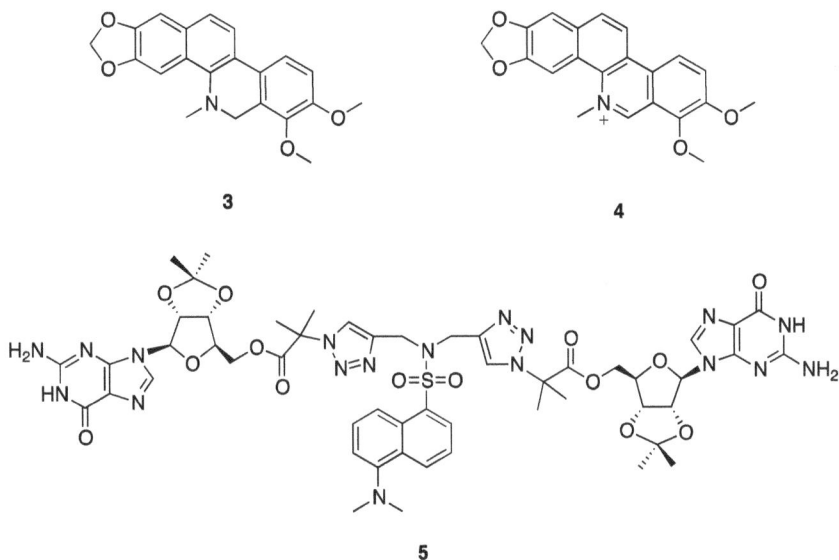

Figure 5.3.   Molecular structures of compounds **3–5**.

consistent with the deactivation of c-MYC promoter, through the stabilization of its G-quadruplex by **5**.

A specific interaction with parallel G-quadruplex DNA was found by Diveshkumar *et al.* when testing compounds **6** and **7** [43] (see Figure 5.4). As depicted in Figure 5.5, both of them are indolylmethyleneindanone-based molecules presenting a positive charge, which, above all, increases their water solubility. The authors, through CD melting experiments, assessed their higher stabilization of the parallel c-MYC and c-KIT G-quadruplexes with respect to both the hybrid telomeric G-quadruplex and duplex DNA. These results were corroborated by several studies such as ESI–MS experiments. The latter are commonly used to assess the occurrence of non-covalent interactions between a G-quadruplex and a small binding molecule and allow determining the affinity constant for the resulting complex [44, 45]. It was therefore found that derivative **6** was a stronger stabilizer of both c-MYC and c-KIT G-quadruplexes than **7**, probably for the presence of the ethyl instead of the propyl side chains of **7** [43]. Notably, by employing MD simulations,

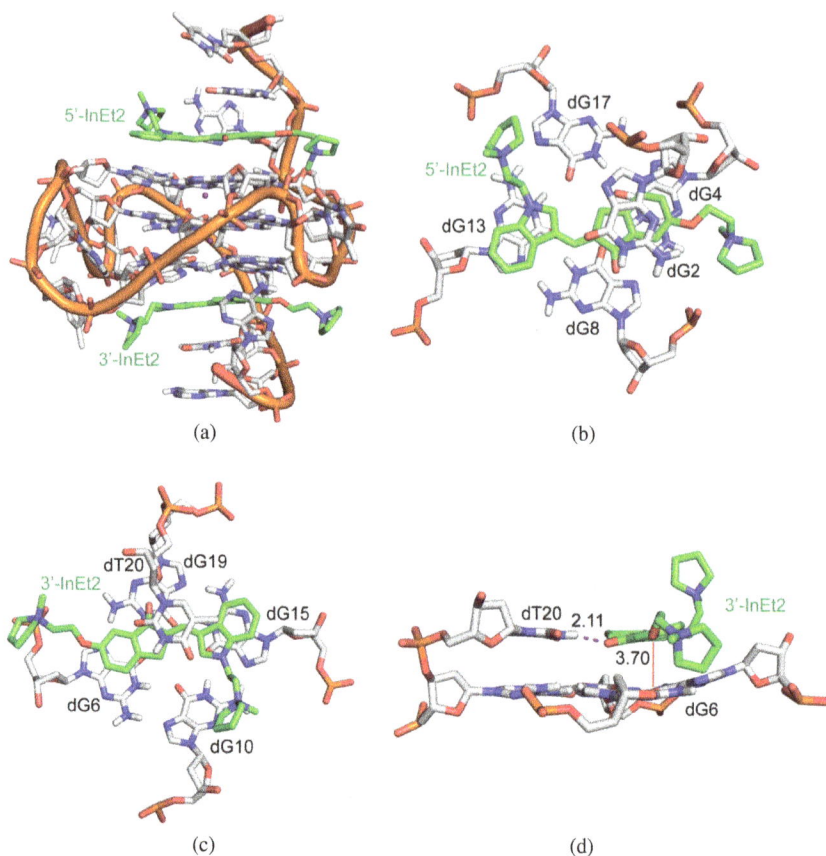

Figure 5.4. MD simulations of **6** with (a) c-MYC G-quadruplex; (b) 5′-quartet and (c, d) 3′-quartet of the G-quadruplex. Adapted with permission from Ref. [43]. Copyright (2016) American Chemical Society.

the authors revealed that the main stabilizing interaction was π-stacking of the two aromatic groups, indole and indanone, with the top and bottom G-quartets and with the flanking bases of c-MYC and c-KIT G-quadruplexes.

This π-stacking was not observed with telomeric G-quadruplex DNA, thus addressing the specific interaction of **6** and **7** with c-MYC and c-KIT rather than with telomeric G-quadruplex DNA to their different topologies. This last evidence was even pointed out by Taq polymerase stop assays, where both molecules showed a low and a high IC_{50} value for c-MYC template and for the telomeric sequence, respectively [43].

Interestingly, other indole-based molecules have been reported in 2016 as promising G-quadruplex stabilizers. As an example, Livendahl *et al.* synthesized several 2,2′-diindolylmethanes derivatives and tested them as G-quadruplex binders by means of spectroscopic and biologic investigations [46]. Among all of the synthesized molecules, **9(a)**, **9(b)** and **10** in Figure 5.5 have shown the best G-quadruplex binding capabilities, especially when compared to the well-known G-quadruplex stabilizer Phen-DC3, **8**. Notably, even these authors found their compounds able to bind more efficiently to certain G-quadruplex sequences, thus allowing them to conclude that upcoming chemical modification of these molecules might enhance their binding selectivity [46].

Moreover, Amato and coworkers recently reported the synthesis of several hydrazone-based compounds as G-quadruplex binders. Among them, derivative **11**, **12** and **13** showed a preferential binding to parallel topologies over human telomeric and duplex DNA, as obtained through CD melting assays [47] (see Figure 5.6).

Molecular docking approaches suggested that the tricyclic diimidazo[1,2-a:1,2-c]pyrimidine core of **12** could easily fill the spaces both in the top and in the bottom faces of c-MYC G-quadruplex. Notably, additional hydrogen bonds and electrostatic attraction reinforced the interactions of **12** with the selected G-quadruplex. Remarkably, the molecule was even found to be the more effective in inhibiting human U2OS and HCT116 cancer cell growth.

Pradeepkumar and coworkers have found a similar binding specificity toward parallel G-quadruplex topologies. Having synthesized three benzothiazole hydrazone of furylbenzamides (**14(a)**, **14(b)** and **14(c)** in Figure 5.7) with different side chains, the authors assessed their G-quadruplex stabilization properties by means of fluorescence and CD melting assays [48]. Interestingly, all of them showed a strong stabilization of parallel G-quadruplex DNA topologies such as those of c-MYC and c-KIT1; on the contrary, the three derivatives presented weaker interactions with antiparallel (h-RAS1) and hybrid (telomeric) G-quadruplexes, as well as with duplex DNA. Notably, derivative **14(b)**, which from the biophysical assays has been recognized as the best of the series, could also inhibit the Taq DNA polymerase of c-MYC template in a micromolar concentration; this value, being about 200-fold lower than the one needed for the telomeric template, confirmed the preference of interaction with parallel

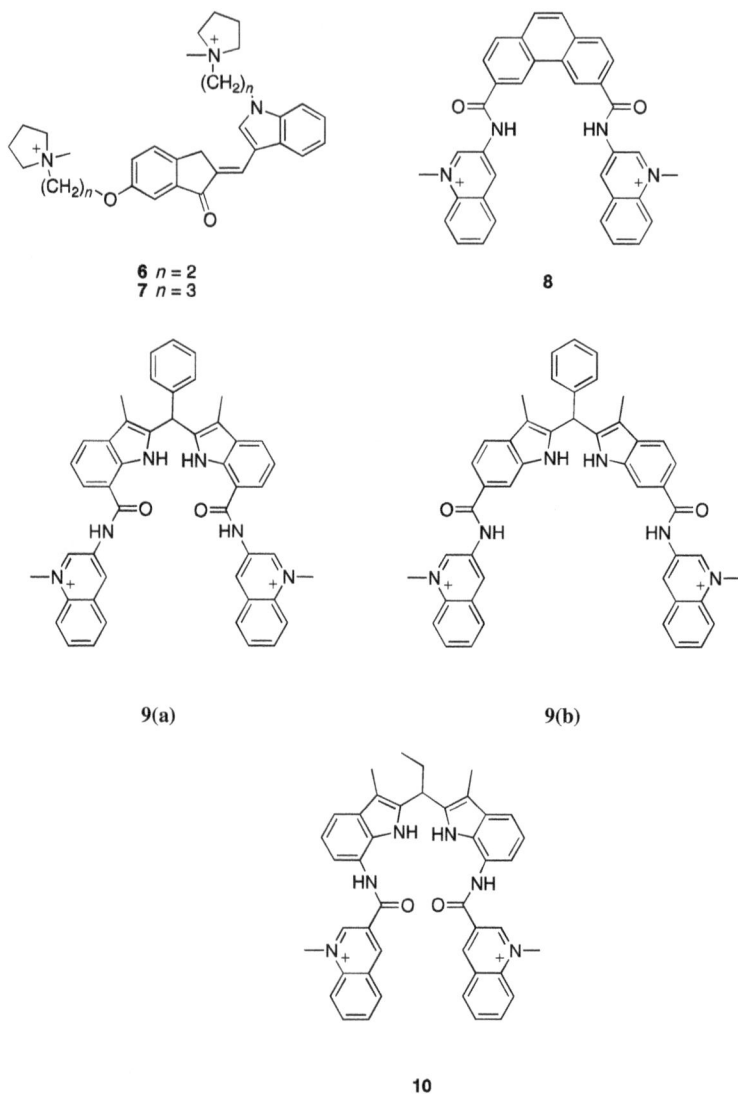

Figure 5.5. Molecular structures of compounds **6–10**.

G-quadruplex topologies. Lastly, computational approaches, used to collect clues about the biding modes of the molecules to c-MYC G-quadruplex DNA, allowed assessing a 1:1 binding stoichiometry in which π-stacking and electrostatic interactions were the main stabilizing forces.

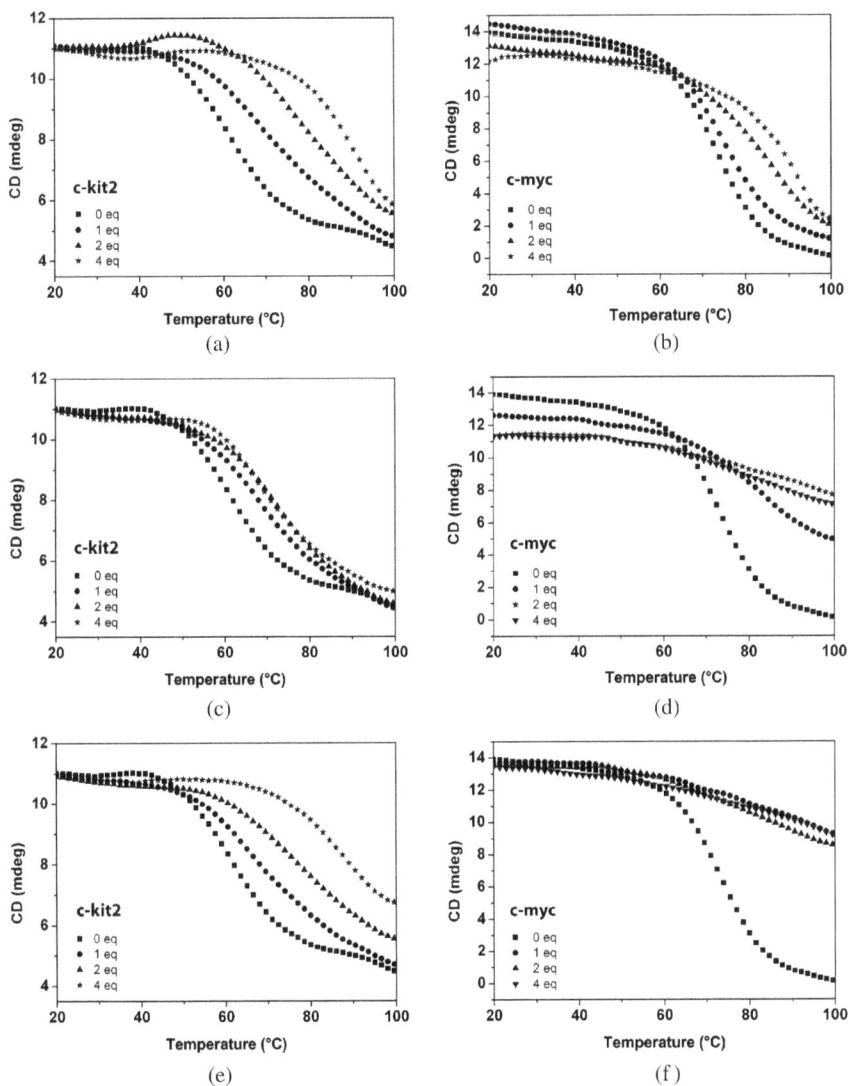

Figure 5.6. CD melting curves of c-KIT (left column) and c-MYC (right column) G-quadruplex DNA in presence of increasing amounts of **11** (a, b), **12** (c, d) and **13** (e, f). Adapted with permission from Ref. [47]. Copyright (2016) American Chemical Society.

R. Bonsignore, E. Trippodo and G. Barone

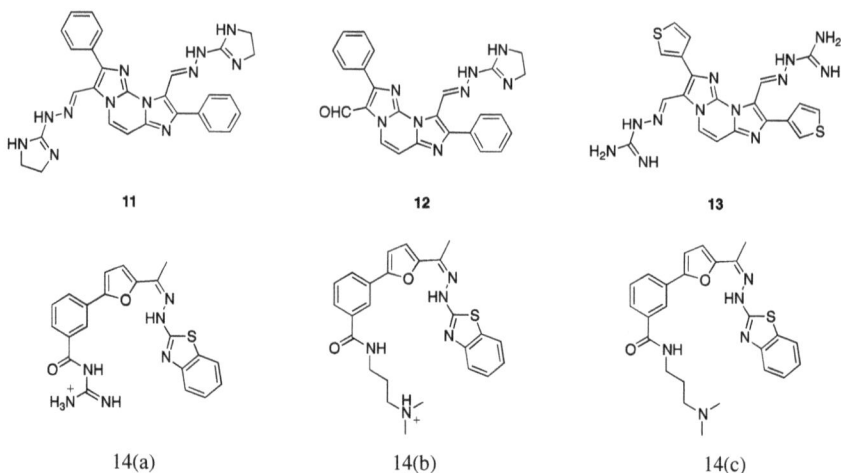

Figure 5.7. Molecular structures of compounds **11–14**.

In the search for c-MYC G-quadruplex selective binders, Jiang and coworkers have reported on the synthesis of novel 4-anilinoquinazoline derivatives [49]. Biophysical studies allowed them to assess that the presence of an aniline group in the quinozaline scaffold, together with two positive charges electrostatically bonding to the loops and groves of the G-quadruplex, improved the binding capabilities of the synthesized molecules, without affecting the duplex DNA structure. Notably, the authors carried out FRET titrations, which allowed them selecting molecule **15** in Figure 5.8 as the most promising derivative of the series. Indeed, in order to carry out a FRET melting assays, a polynucleotide is labeled with a fluorophore on one of its ends and a fluorescence quencher on the opposite end. Their proximity allows quenching the fluorescence of the labeled polynucleotide. On the contrary, when temperature is increased the denaturation of the DNA occurs leading to the separation of the bases and therefore moving apart the fluorophore and the quencher. For such a reason, an increase of the fluorescence signal can be recorded and allows determining the temperature of melting (T_m) of the polynucleotide. Once a molecule is stabilizing the duplex or the G-quadruplex DNA, more energy is needed to break the hydrogen bonds between the bases and therefore a shift of the T_m toward higher temperature is expected.

15

16

17

18

19

20

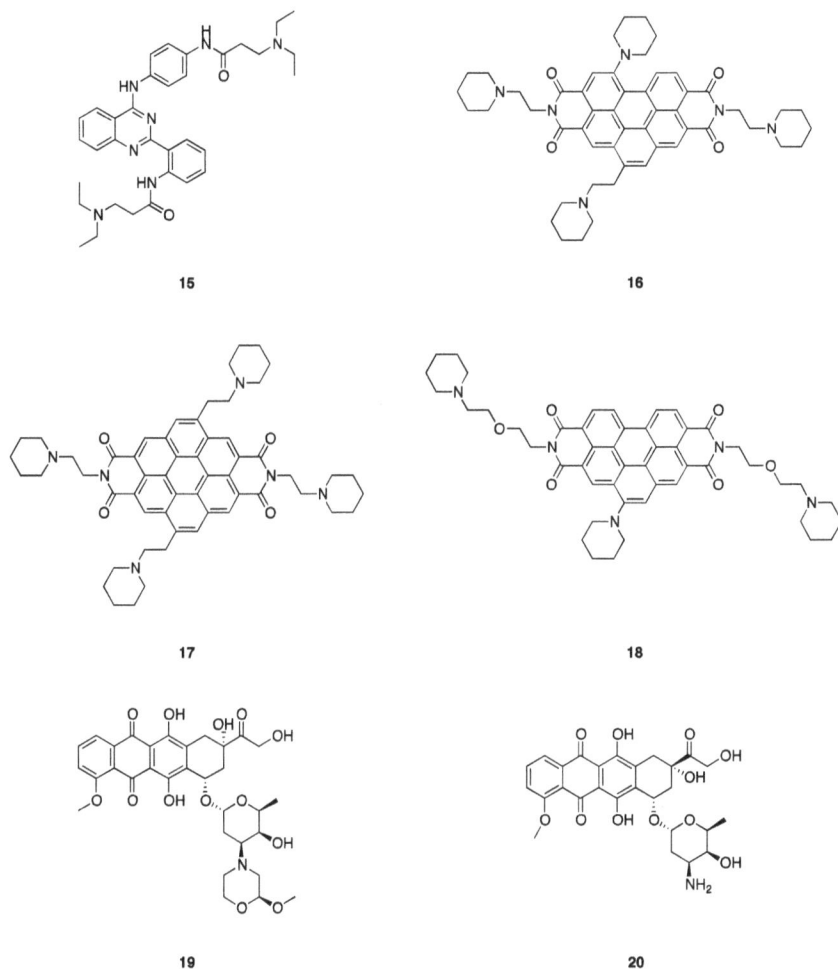

Figure 5.8. Molecular structures of compounds **15–20**.

In particular, derivative **15** has shown an about 20°C increase in the T_m of the c-MYC G-quadruplex DNA, pointing out its strong binding to the latter. The authors have recognized this stabilization as the most probably cause for the down-regulation of the expression of the protein c-MYC when HeLa cells were exposed to **15**. Further investigations have even proved that the molecule did not affect the growth of normal primary cultured mouse mesangial cells, despite having an antiproliferative effect on HeLa cell lines, with their arrest in G0/G1 cell cycle phase.

Micheli *et al.* has recently deepened the studies on derivative **16** (Figure 5.8) [50] — addressed in a past work as a human telomeric G-quadruplex stabilizer [51] — investigating on its ability to stabilize also c-MYC and BCL-2 G-quadruplex DNA. The molecule is structurally related to both compounds **17** and **18**, which for such a reason were included in the assays carried out. Collectively, biophysical experiments such as ESI–MS, FRET and CD titrations have shown **16** and **17** have comparable G-quadruplex stabilizing capabilities, outperforming the other related molecule, **18**. As already mentioned for several compounds reported in this paragraph, authors have shown that **16** could interact with G-quadruplex DNA through π-stacking to the ending G-quartets, even though no selectivity of interaction with the c-MYC and BCL-2 could be observed. Lastly, qRT-PCR and western blotting analysis showed that **16** was the most effective molecule in reducing the protein levels of both of the genes. Consequently, the authors concluded that the anti-cancer capabilities of the derivative could not be merely related to its interaction with human telomeric G-quadruplex DNA but also with other G-quadruplexes. Notably, since **16** has shown to be a better *in vitro* antiproliferative derivative than compounds **17** and **18**, it was considered a suitable candidate for developing novel and more specific G-quadruplex binders.

Lastly, the anti-cancer capabilities of anthracyclines have been widely known to be related to a multiple mechanism of action, which can, directly or indirectly, even involve the duplex DNA [52]. Nemorubicin and doxorubicin, respectively **19** and **20** in Figure 5.7, are indeed known to be DNA-intercalators. Nevertheless, their molecular structures meet almost all the requirements to interact also with G-quadruplex DNA. For such a reason, Scaglioni and coworkers performed NMR and ESI–MS experiments with an oligonucleotide containing the TTAGGGT repeated sequence, typical of the human telomeres, in order to assess the G-quadruplex binding capabilities of both **19** and **20** [52]. Not surprisingly, the authors found that similar interactions with the telomeric G-quadruplex DNA could be achieved in presence of the two molecules and structural evidences of the interactions with the polynucleotide could be collected from their experiments. In particular **19**, which showed the best binding capabilities, has been found to be able to intercalate between the A3 of the loop and the G-tetrad, in the 5′-end of the G-quadruplex; at the same time, the molecule was also capable to intercalate between the G6 and the T6, in the 3′-end. A similar 2:1 stoichiometry of interaction was also found when the authors tested **19** with c-MYC sequence, by means of the same spectroscopic approaches. Nonetheless, they pointed out in the latter case that the interaction with the 5′-end should be much more favorable, because seven π-interactions could be concomitantly achieved. In conclusion, the findings of the authors clearly demonstrated the capabilities of **19** and **20** to interact also with G-quadruplex DNA therefore allowing them to infer that their anti-cancer capabilities could be also due to these interactions.

5.4.2 Targeting G-quadruplex DNA with metal complexes

Ruthenium based molecules have been widely investigated in past decades as promising anti-cancer candidates: indeed, displaying a remarkable *in vitro* anti-cancer activity, this metal has gained a role as a possible alternative to platinum based drugs [53]. Notably, complexes such as NAMI A and KP1019 (**21** and **22** in Figure 5.9) have also undergone phase I and II clinical trials [54–56].

He *et al.*, in 2016 synthesized three octahedral Ru^{II} complexes with aromatic planar ligands: among them, complex **24** (Figure 5.9) bearing two ammonia groups along the two z-axial position, showed the best DNA binding capabilities [57]. By means of biophysical assays, the authors proved the occurrence of significant topological selectivity of **24** towards both the antiparallel telomeric G-quadruplexes investigated (hybrid and basket-type) rather than towards the parallel conformation, represented by c-MYC. The thermal stability and the CD spectra of c-MYC G-quadruplex were, indeed, much less altered in presence of the derivative. Moreover, the comparison between the two antiparallel telomeric G-quadruplexs binding with **24**, indicated a higher structural fitting with the basket-type rather than with the hybrid telomeric G-quadruplex. Computational studies confirmed the occurrence of π-stacking of **24** with the G-quartets. Notably, the presence of the ammonia groups allowed achieving unique extra stabilizing effects: in fact NH$_3$ could insert into the central cationic channel and interact with the bases through hydrogen bonds.

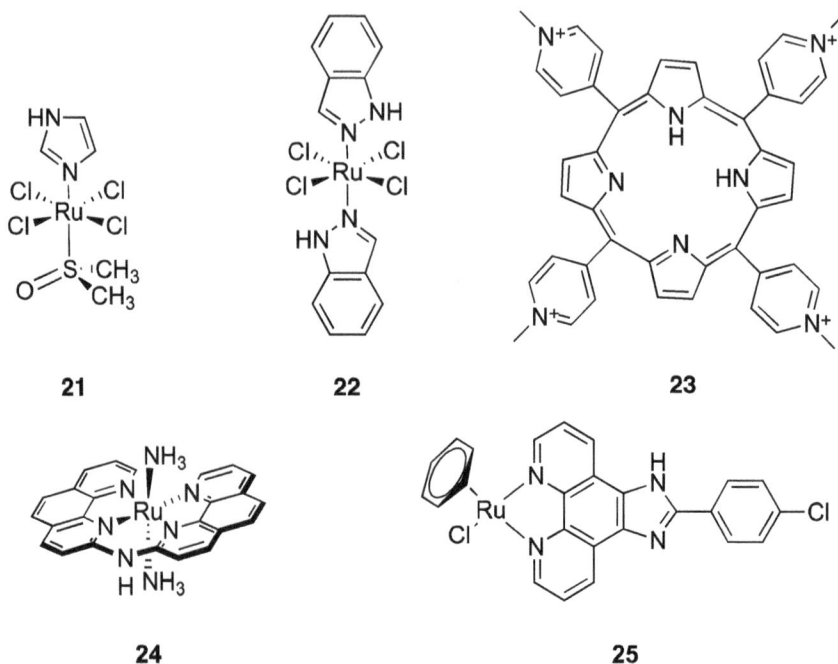

Figure 5.9. Molecular structures of compounds **21–25**.

Moreover, complex **24** exhibited an efficient inhibition of the telom-erase activity, at a much lower concentration than that needed for BRACO-19 (**1**) and TMPyP4 (**23**) [58], well-known G-quadruplex bind-ers. Lastly, its cytotoxicity was investigated against HeLa, MDA–MB231, HT 1080 and MCF-7 cancer cell lines, showing that **24** was about 2 times more potent than TMPyP4. Remarkably, such Ru^{II} complex was even less cytotoxic towards normal cells [57].

Another interesting study has been carried out by Wu and coworkers, synthesizing a series of arene Ru complexes with different halides substi-tuting an hydrogen atom, and studying the influence of this substituent on c-MYC G-quadruplex binding and its biological consequences [59]. Results, supported by computational studies, have demonstrated that all complexes are able to bind to the grooves of the G-quadruplex but derivative **25** in Figure 5.9, bearing a chloride as substituent, was the best of the series, showing a higher affinity toward the G-quadruplex in the biophysical assays. Indeed, atomic radii seemed to play a key role in the groove-binding process. In particular, CD titrations showed remarkable variations on c-MYC spectra in presence of increasing amounts of **25**, which are consistent with conformational changes occurring to G-quadruplex DNA's secondary structure.

Wu *et al.* also used the FITC-gelatin invasion assay to assess the anti-cancer capabilities of the arene–Ru complexes. Indeed cells with high invasion rates can form protrusion of the plasma membrane (namely invadopodia) and release matrix metalloproteinases. The latter degrade the fluorescent gelatin and therefore the appearance of dark holes or of non-fluorescent areas occurs. MDA–MB231 cells, without treatment with complex **25**, formed lots of black holes in FITC-gelatin, whereas after their exposure to **25** the dark areas decreased (Figure 5.10), proving a marked inhibition of invasion of MDA–MB231 cells [59].

The toxicity of the molecule has been tested using zebrafish embryos. It is worth reminding that zebrafish DNA possesses a high homology with human DNA and therefore they are considered reliable models for human

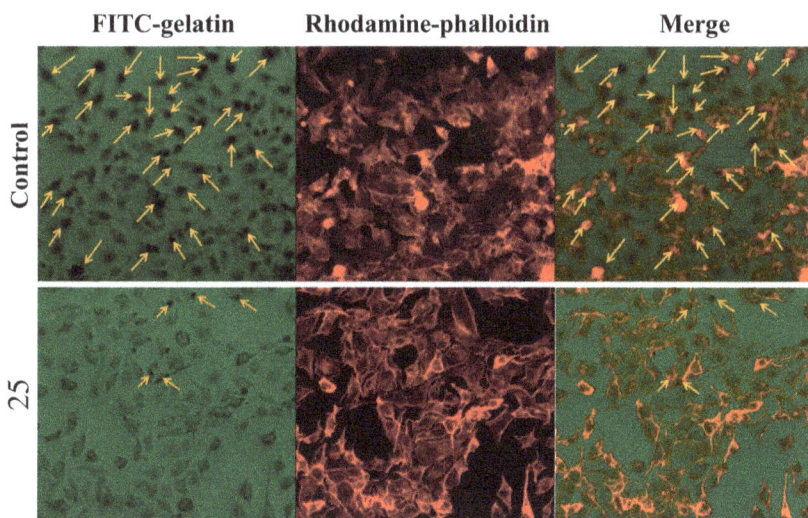

Figure 5.10. FITC gelatin assays of derivative **25**. Adapted with permission from Ref. [59]. Copyright (2016) American Chemical Society.

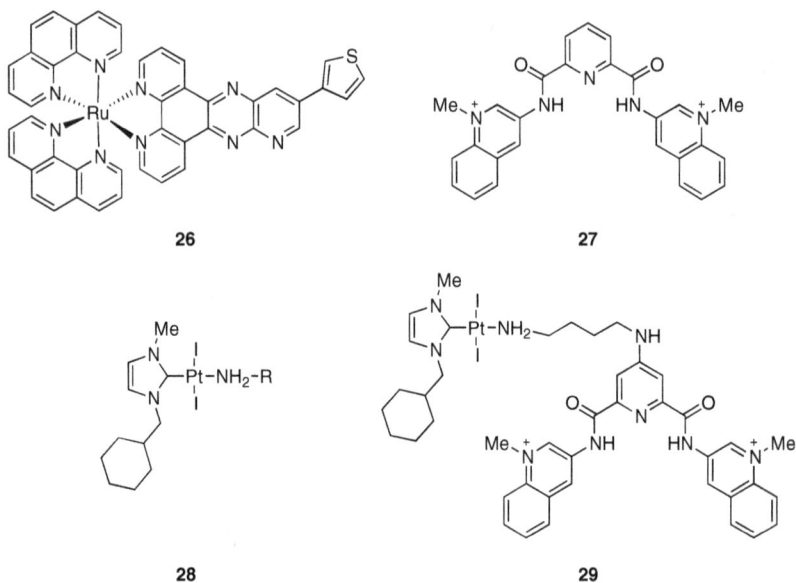

Figure 5.11. Molecular structures of compounds **26–29**.

DNA-binders. All the zebrafish embryos, without treatment with complex **25**, developed into *juvenile* zebrafish; interestingly the development of the embryos was even achieved after exposure to increasing concentrations of **25**, suggesting its low toxicity for the zebrafish embryos [59].

Interestingly, Wang and coworkers synthesized and characterized two similar ruthenium(II) complexes, an example of which is **26** in Figure 5.11 [60]. By means of biophysical assays they proved that both complexes were able to strongly bind and stabilize human telomeric G-quadruplex DNA. In particular, Wang *et al.* carried out a Fluorescent Intercalator Displacement (FID) assay, based on the competitive binding with DNA by synthetic molecules and Thiazole Orange (TO). Indeed, TO, an organic molecule with a high affinity towards DNA, displays a fluorescence signal when it is bound to the polynucleotide; on the contrary in presence of a competitive ligand for G-quadruplex DNA, its displacement occurs and therefore a loss of the fluorescence of TO is achieved.

The two Ru^{II} complexes were able to displace TO at low concentrations, resulting stronger G-quadruplex binders than TO. Moreover both of them, but in particular **26**, displayed a preferential binding to human telomeric G-quadruplex rather than to duplex DNA [60].

Platinum complexes have been widely studied as anti-cancer agent since the discovery of cisplatin. This complex, able to bind covalently B-DNA, is still used in anti-cancer treatments and, for this reason, is often included in binding assays as a comparison.

Betzer and coworkers conjugated the N-heterocyclic carbene platinum complex **27**, which had already showed anti-tumor properties, with the G-quadruplex binder PDC **28** [61]. The synthesized complex, **29**, displayed enhanced properties with respect to **27** and **28** alone, with a binding process resulting from the cooperation of the two components of the molecule. Indeed, **29** was able to bind preferentially and irreversibly human telomeric G-quadruplex sequences with a stronger affinity than **28**. Moreover, by means of biophysical assays Betzer *et al.* proved that **29** bound covalently G-quadruplex DNA with a different platination process. By isolating platination products it was possible to establish the metallic coordination of **29** with adenines A7 and A19 differently from the complex **27** that was coordinated with adenines A7 and A13. This result allowed deducing that the binding of **29** was driven by the stacking of the PDC moiety **28** onto the G4-structure. Moreover, biological assays proved that the conjugated complex **29** induced an increased loss of TRF2, a telomeric protein essential for the maintenance of telomeres [62], compared to **27** and **28** alone [61].

Furthermore, Gama and coworkers synthesized a series of anthracene–terpyridine Pt(II) and Cu(II) complexes containing different linkers [63]. By biophysical assays these derivatives showed affinity and selectivity for human telomeric and c-MYC G-quadruplexes over duplex DNA. Notably, the metal-free ligands did not display any binding capabilities towards duplex and G-quadruplex DNA, highlighting the importance of the metallic center. Interestingly, the thermal stabilization of the G-quadruplexes induced by these compounds was higher in the case of platinum(II) complexes than of the copper(II) ones. Furthermore, complexes with a longer linker, for example **30**, showed enhanced interaction with G-quadruplex forming sequences [63].

Moreover, Duskova and coworkers studied the interaction of a series of Cu(II) complexes of carbohydrate ligands, whose general structure, **31**,

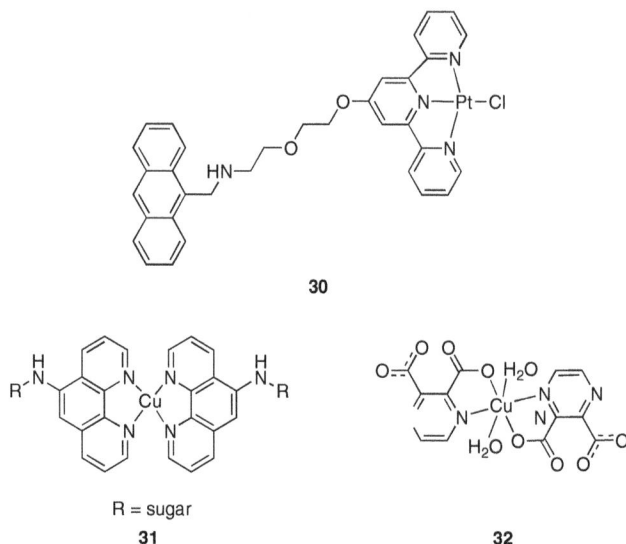

Figure 5.12. Molecular structures of compounds **30–32**.

is reported in Figure 5.12, with human telomeric G-quadruplex DNA [64]. Interestingly, among the complexes containing carbohydrate units of the D-series, the glucose and mannose derivatives were the most able to displace TO in the G-quadruplex FID assay and showed higher selectivity for G-quadruplex over duplex DNA under potassium ionic conditions. On the contrary, for the L-series containing complexes the authors found the best derivative in the rhamnose complex, possessing the higher binding affinity and selectivity. It also exhibited the most drastic modification of the typical CD spectra of G-quadruplex DNA, suggesting the occurrence of a considerable modification of the secondary structure of the polynucleotide. Interestingly, this rhamnose derivate outperformed all of the other Cu(II) complexes. In conclusion, the authors were able to suppose an interaction mode of the synthesized compounds which involves their binding with the loops and grooves of the polynucleotide, despite their π-stacking, as supported by TO displacement, still plays an important role [64].

Concerning other copper(II) complexes, Housaindokht reported about a novel compound, **32**, and, by means of biophysical assays, studied its

G-quadruplex binding capabilities [65]. The UV–Vis spectra of the complex, recorded in presence of increasing amounts of the polynucleotide, showed a hypochromic effect, typical of the interaction with DNA. Moreover, MD simulations confirmed the ability of **32** to stack on the quartets and stabilize the structure. Thermodynamic parameters calculated from the biophysical assays suggested that the binding process was driven by the entropy and displayed the key role of hydrophobic forces in the interaction of **32** with G-quadruplex DNA [65]. Moreover the negative values of ΔG indicated the spontaneity of the binding process.

Lastly, for their extended π-system, Schiff base derivatives such as Salen-like and Salphen-like transition metal complexes have been widely investigated for their DNA-binding capabilities. A series of publications has indeed pointed out that these molecules act as DNA-intercalators and as G-quadruplex stabilizers [66–70]. In such a context, Bonsignore *et al.* reported the synthesis of two novel zinc(II) and nickel(II) Salphen-like compounds of which the nickel(II) complex, **33** in Figure 5.13, showed to be the only one capable to interact with telomeric G-quadruplex [71]. By combining UV–Vis and CD spectroscopies, the authors assessed the preferential binding of this derivative with human telomeric G-quadruplex

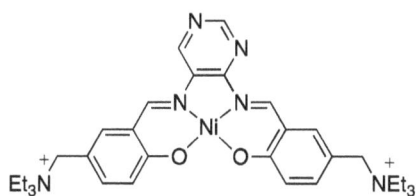

33

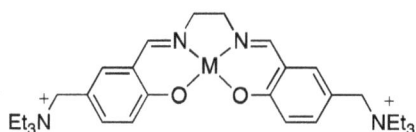

34 M = Ni
35 M = Cu

Figure 5.13. Molecular structures of compounds **33–35**.

with respect to duplex DNA. Notably, MD simulations pointed out that **33** found its equilibrium when the Ni^{II} atom was almost aligned with the potassium channel. Therefore, together with the presence of π–π and electrostatic interactions, the results confirmed that the presence of a metal center improves the stabilization of the molecule onto the external face of the G-quadruplex.

Similarly, Terenzi *et al.* have reported the synthesis of two nickel(II) and copper(II) complexes of a Salen-like ligand (**34** and **35** in

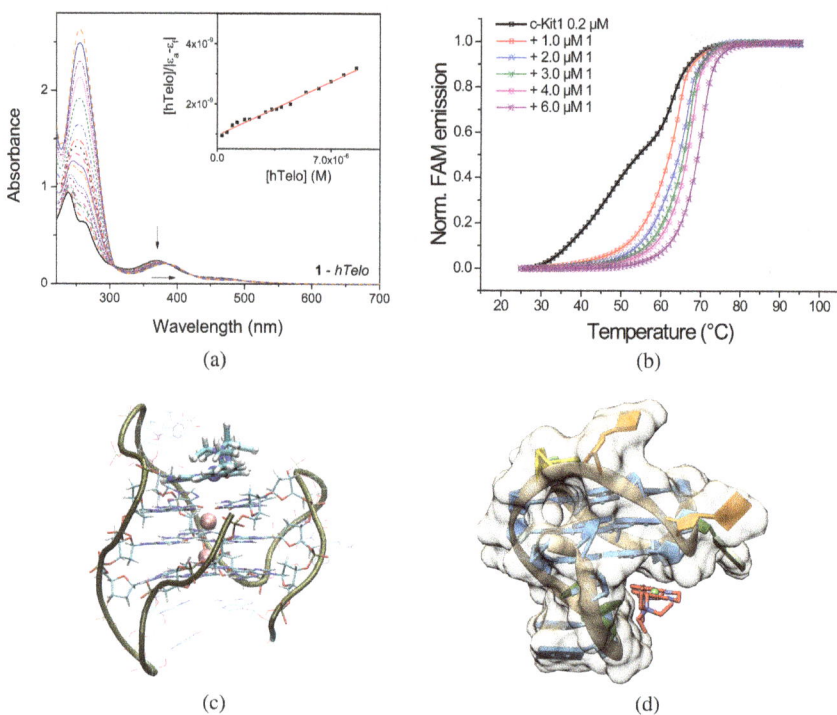

Figure 5.14. (a) UV–Vis titration of a nickel(II) complex, **33**, in presence of human telomeric G-quadruplex DNA; (b) FRET melting assays of c-KIT G-quadruplex DNA in presence of the derivative **34**; (c) molecular dynamic simulation of **33** with human telomeric G-quadruplex DNA; (d) docking simulation of **34** with c-KIT G-quadruplex DNA. (a), (c): Adapted with permission from Ref. [71]. Copyright (2016) Elsevier; (b), (d): Adapted with permission from Ref. [72]. Copyright (2016) Published by The Royal Society of Chemistry.

Figure 5.13) [72]. Differently from derivative **33**, both nickel(II), **34**, and copper(II), **35**, complexes are not characterized by the presence of an aromatic ring on the N–N bridge. Spectrophotometric titrations revealed that **34** possesses the best G-quadruplex stabilizing capabilities, especially toward the c-KIT sequence. Moreover, none of the complexes interacts with duplex DNA. This last result clearly pointed out that both derivatives were capable to discriminate between the double-helical and the G-quadruplex secondary structure of the polynucleotide, therefore acting as selective G-quadruplex binders. Notably, by means of molecular modeling approaches the authors realized that the mechanism of action of **34** was different from the known G-quadruplex top/end stackers. Indeed, the computational investigation points out the occurrence of groove binding of **34** into the characteristic cleft of c-KIT G-quadruplex (Figure 5.14), thus explaining its preferential binding towards the latter G-quadruplex rather than any other and duplex DNA. Lastly, **34** in presence of a lipophilic carrier showed a stronger cytotoxic activity than **35**, reproducing *in vitro* the same results achieved with G-quadruplex affinity studies [72].

5.5 Conclusion and Perspectives

The localization in human telomeres and in oncogene promoters of G-quadruplex DNA has attracted the attention of many researchers seeking for selective anti-cancer molecules. Notably, these secondary structures have been found to be involved in other regulatory cellular processes such as DNA replication or translation of mRNAs [2,3]. For example, they have been found in 5′-untranslated regions (5′-UTR) of mRNAs where they play a role discriminating whether the binding of the polynucleotide with the ribosome will occur or not, thus regulating the transcription of a protein. Interestingly, recent structural studies have allowed stating that also Zika's virus 3′-UTR RNA might adopt stable parallel G-quadruplex structures, opening the way to new possible medical treatments [73].

In 2016, a large amount of G-quadruplex agents as anti-cancer scaffolds has been reported in the literature. Although these have not yet been tested *in vivo*, they represent a valuable database of promising candidates that might be chemically modified to improve their G-quadruplex binding capabilities and enhance their biological activities. In this context, it is to be expected that an important role in designing novel G-quadruplex binders will be increasingly played in the near future by computational methods. Overall, several steps are still needed to take before finding a G-quadruplex binder in the shelves of a pharmacy. Nonetheless, the discoveries reported so far allow being optimistic and represent a milestone in the search of this aim.

References

1. Neidle, S. (2011). *Therapeutic Applications of Quadruplex Nucleic Acids.* Academic Press Inc: London, Waltham, MA.
2. Neidle, S. (2016). Quadruplex nucleic acids as novel therapeutic targets. *J. Med. Chem.* **59**, 5987–6011.
3. Rhodes, D., and Lipps, H.J. (2015). G-quadruplexes and their regulatory roles in biology. *Nucleic Acids Res.* **43**, 8627–8637.
4. Bang, I. (1910). Untersuchungen über die guanylsäre. *Biochem. Z.* **26**, 293–311.
5. Ralph, R.K., Connors, W.J., and Khorana, H.G. (1962). Secondary structure and aggregation in deoxyguanosine oligonucleotides. *J. Am. Chem. Soc.* **84**, 2265–2266.
6. Gellert, M., Lipsett, M.N., and Davies, D.R. (1962). Helix formation by guanylic acid. *Proc. Natl. Acad. Sci. USA.* **48**, 2013–2018.
7. Sen, D., and Gilbert, W. (1990). A sodium-potassium switch in the formation of four-stranded G4-DNA. *Nature.* **344**, 410–414.
8. Neidle, S., and Balasubramanian, S. (2006). *Quadruplex Nucleic Acids.* Royal Society of Chemistry Publishing: Cambridge, UK.
9. Düchler, M. (2012). G-quadruplexes: Targets and tools in anticancer drug design. *J. Drug Target.* **20**, 389–400.
10. Schaffitzel, C., Berger, I., Postberg, J., Hanes, J., Lipps, H.J., and Plückthun, A. (2001). *In vitro* generated antibodies specific for telomeric guanine-quadruplex DNA react with Stylonychia lemnae macronuclei. *Proc. Natl. Acad. Sci. USA.* **98**, 8572–8577.
11. Biffi, G., Tannahill, D., McCafferty, J., and Balasubramanian, S. (2013). Quantitative visualization of DNA G-quadruplex structures in human cells. *Nat. Chem.* **5**, 182–186.
12. Lam, E.Y.N., Beraldi, D., Tannahill, D., and Balasubramanian, S. (2013). G-quadruplex structures are stable and detectable in human genomic DNA. *Nat. Commun.* **4**, 1796.

13. Huppert, J.L., and Balasubramanian, S. (2005). Prevalence of quadruplexes in the human genome. *Nucleic Acids Res.* **33**, 2908–2916.
14. Larsen, A., and Weintraub, H. (1982). An altered DNA conformation detected by S1 nuclease occurs at specific regions in active chick globin chromatin. *Cell.* **29**, 609–622.
15. Rustighi, A., Tessari, M.A., Vascotto, F., Sgarra, R., Giancotti, V., and Manfioletti, G. (2002). A polypyrimidine/polypurine tract within the Hmga2 minimal promoter: A common feature of many growth-related genes. *Biochemistry.* **41**, 1229–1240.
16. Sun, D. Guo, K., Rusche, J.J., and Hurley, L.H. (2005). Facilitation of a structural transition in the polypurine/polypyrimidine tract within the proximal promoter region of the human VEGF gene by the presence of potassium and G-quadruplex-interactive agents. *Nucleic Acids Res.* **33**, 6070–6080.
17. Simonsson, T., Pecinka, P., and Kubista, M. (1998). DNA tetraplex formation in the control region of c-myc. *Nucleic Acids Res.* **26**, 1167–1172.
18. Siddiqui-Jain, A., Grand, C.L., Bearss, D.J., and Hurley, L.H. (2002). Direct evidence for a G-quadruplex in a promoter region and its targeting with a small molecule to repress c-MYC transcription. *Proc. Natl. Acad. Sci. USA.* **99**, 11593–11598.
19. De Armond, R., Wood, S., Sun, D., Hurley, L.H., and Ebbinghaus, S.W. (2005). Evidence for the presence of a guanine quadruplex forming region within a polypurine tract of the hypoxia inducible factor 1alpha promoter. *Biochemistry.* **44**, 16341–16350.
20. Dai, J., Chen, D., Jones, R.A., Hurley, L.H., and Yang, D. (2006). NMR solution structure of the major G-quadruplex structure formed in the human BCL2 promoter region. *Nucleic Acids Res.* **34**, 5133–5144.
21. Dexheimer, T.S., Sun, D., and Hurley, L.H. (2006). Deconvoluting the structural and drug-recognition complexity of the G-quadruplex-forming region upstream of the bcl-2 P1 promoter. *J. Am. Chem. Soc.* **128**, 5404–5415.
22. Dai, J., Dexheimer, T.S., Chen, D., Carver, M., Ambrus, A., Jones, R.A., and Yang, D. (2006). An intramolecular G-quadruplex structure with mixed parallel/antiparallel G-strands formed in the human BCL-2 promoter region in solution. *J. Am. Chem. Soc.* **128**, 1096–1098.
23. Yafe, A., Etzioni, S., Weisman-Shomer, P., and Fry, M. (2005). Formation and properties of hairpin and tetraplex structures of guanine-rich regulatory sequences of muscle-specific genes. *Nucleic Acids Res.* **33**, 2887–2900.
24. Cogoi, S., Quadrifoglio, F., and Xodo, L.E. (2004). G-rich oligonucleotide inhibits the binding of a nuclear protein to the Ki-ras promoter and strongly reduces cell growth in human carcinoma pancreatic cells. *Biochemistry.* **43**, 2512–2523.

25. Rankin, S., Reszka, A.P., Huppert, J., Zloh, M., Parkinson, G.N., Todd, A.K., Ladame, S., Balasubramanian, S., and Neidle, S. (2005). Putative DNA quadruplex formation within the human c-kit oncogene. *J. Am. Chem. Soc.* **127**, 10584–10589.

26. Fernando, H., Reszka, A.P., Huppert, J., Ladame, S., Rankin, S., Venkitaraman, A.R., Neidle, S., and Balasubramanian, S. (2006). A conserved quadruplex motif located in a transcription activation site of the human c-kit oncogene. *Biochemistry.* **45**, 7854–7860.

27. Guo, K., Pourpak, A., Beetz-Rogers, K., Gokhale, V., Sun, D., and Hurley, L.H. (2007). Formation of pseudosymmetrical G-quadruplex and i-motif structures in the proximal promoter region of the RET oncogene. *J. Am. Chem. Soc.* **129**, 10220–10228.

28. Qin, Y., Rezler, E.M., Gokhale, V., Sun, D., and Hurley, L.H. (2007). Characterization of the G-quadruplexes in the duplex nuclease hypersensitive element of the PDGF-A promoter and modulation of PDGF-A promoter activity by TMPyP4. *Nucleic Acids Res.* **35**, 7698–7713.

29. Cech, T.R. (2004). Beginning to understand the end of the chromosome. *Cell.* **116**, 273–279.

30. Aisner, D.L., Wright, W.E., and Shay, J.W. (2002). Telomerase regulation: Not just flipping the switch. *Curr. Opin. Genet. Dev.* **12**, 80–85.

31. Lauria, A., Terenzi, A., Bartolotta, R., Bonsignore, R., Perricone, U., Tutone, M., Martorana, A., Barone, G., and Almerico, A.M. (2014). Does ligand symmetry play a role in the stabilization of DNA G-quadruplex host-guest complexes? *Curr. Med. Chem.* **21**, 2665–2690.

32. Heaphy, C.M., Subhawong, A.P., Hong, S.-M., Goggins, M.G., Montgomery, E.A., Gabrielson, E., Netto, G.J., Epstein, J.I., Lotan, T.L., Westra, W.H., Shih, I.-M., Iacobuzio-Donahue, C.A., Maitra, A., Li, Q.K., Eberhart, C.G., Taube, J.M., Rakheja, D., Kurman, R.J., Wu, T.C., Roden, R.B., Argani, P., De Marzo, A.M., Terracciano, L., Torbenson, M., and Meeker, A.K. (2011). Prevalence of the alternative lengthening of telomeres telomere maintenance mechanism in human cancer subtypes. *Am. J. Pathol.* **179**, 1608–1615.

33. Burger, A.M., Dai, F., Schultes, C.M., Reszka, A.P., Moore, M.J., Double, J.A., and Neidle, S. (2005). The G-quadruplex-interactive molecule BRACO-19 inhibits tumor growth, consistent with telomere targeting and interference with telomerase function. *Cancer Res.* **65**, 1489–1496.

34. Drygin, D., Siddiqui-Jain, A., O'Brien, S., Schwaebe, M., Lin, A., Bliesath, J., Ho, C.B., Proffitt, C., Trent, K., Whitten, J.P., Lim, J.K.C., Von Hoff, D., Anderes, K., and Rice, W.G. (2009). Anticancer activity of CX-3543: A direct inhibitor of rRNA biogenesis. *Cancer Res.* **69**, 7653–7661.

35. Šimánek, V. (1985). Benzophenanthridine alkaloids. In A. Brossi (ed.), *Alkaloids Chem. Pharmacol.*, Academic Press, London, pp. 185–240.

36. Krane, B.D., Fagbule, M.O., Shamma, M., and Gözler, B. (1984). The Benzophenanthridine Alkaloids. *J. Nat. Prod.* **47**, 1–43.

37. Collie, G.W., and Parkinson, G.N. (2011). The application of DNA and RNA G-quadruplexes to therapeutic medicines. *Chem. Soc. Rev.* **40**, 5867–5892.

38. Monchaud, D., and Teulade-Fichou, M.-P. (2008). A hitchhiker's guide to G-quadruplex ligands. *Org. Biomol. Chem.* **6**, 627–636.

39. Ghosh, S., and Dasgupta, D. (2015). Quadruplex forming promoter region of c-myc oncogene as a potential target for a telomerase inhibitory plant alkaloid, chelerythrine. *Biochem. Biophys. Res. Commun.* **459**, 75–80.

40. Cui, X., Lin, S., and Yuan, G. (2012). Spectroscopic probing of recognition of the G-quadruplex in c-kit promoter by small-molecule natural products. *Int. J. Biol. Macromol.* **50**, 996–1001.

41. Malhotra, R., Rarhi, C., Diveshkumar, K.V., Barik, R., D'cunha, R., Dhar, P., Kundu, M., Chattopadhyay, S., Roy, S., Basu, S., Pradeepkumar, P.I., and Hajra, S. (2016). Dihydrochelerythrine and its derivatives: Synthesis and their application as potential G-quadruplex DNA stabilizing agents. *Bioorg. Med. Chem.* **24**, 2887–2896.

42. Pavan Kumar, Y., Saha, P., Saha, D., Bessi, I., Schwalbe, H., Chowdhury, S., and Dash, J. (2016). Fluorescent dansyl-guanosine conjugates that bind *c-MYC* promoter G-quadruplex and downregulate *c-MYC* expression. *ChemBioChem.* **17**, 388–393.

43. Diveshkumar, K.V., Sakrikar, S., Rosu, F., Harikrishna, S., Gabelica, V., and Pradeepkumar, P.I. (2016). Specific stabilization of c-MYC and c-KIT G-quadruplex DNA structures by indolylmethyleneindanone scaffolds. *Biochemistry.* **55**, 3571–3585.

44. Gabelica, V. (2010). Determination of equilibrium association constants of ligand-DNA complexes by electrospray mass spectrometry. *Methods Mol. Biol.* **613**, 89–101.

45. Marchand, A., Granzhan, A., Iida, K., Tsushima, Y., Ma, Y., Nagasawa, K., Teulade-Fichou, M.-P., and Gabelica, V. (2015). Ligand-induced conformational changes with cation ejection upon binding to human telomeric DNA G-quadruplexes. *J. Am. Chem. Soc.* **137**, 750–756.

46. Livendahl, M., Jamroskovic, J., Ivanova, S., Demirel, P., Sabouri, N., and Chorell, E. (2016). Design and synthesis of 2,2′-diindolylmethanes to selectively target certain G-quadruplex DNA structures. *Chem. Eur. J.* **22**, 13004–13009.

47. Amato, J., Morigi, R., Pagano, B., Pagano, A., Ohnmacht, S., De Magis, A., Tiang, Y.-P., Capranico, G., Locatelli, A., Graziadio, A., Leoni, A., Rambaldi, M., Novellino, E., Neidle, S., and Randazzo, A. (2016). Toward the development of specific G-quadruplex binders: Synthesis, biophysical, and biological studies of new hydrazone derivatives. *J. Med. Chem.* **59**, 5706–5720.

48. Pany, S.P.P., Bommisetti, P., Diveshkumar, K.V., and Pradeepkumar, P.I. (2016). Benzothiazole hydrazones of furylbenzamides preferentially stabilize c-MYC and c-KIT1 promoter G-quadruplex DNAs. *Org. Biomol. Chem.* **14**, 5779–5793.

49. Jiang, Y., Chen, A.-C., Kuang, G.-T., Wang, S.-K., Ou, T.-M., Tan, J.-H., Li, D., and Huang, Z.-S. (2016). Design, synthesis and biological evaluation of 4-anilinoquinazoline derivatives as new c-myc G-quadruplex ligands. *Eur. J. Med. Chem.* **122**, 264–279.

50. Micheli, E., Altieri, A., Cianni, L., Cingolani, C., Iachettini, S., Bianco, A., Leonetti, C., Cacchione, S., Biroccio, A., Franceschin, M., and Rizzo, A. (2016). Perylene and coronene derivatives binding to G-rich promoter oncogene sequences efficiently reduce their expression in cancer cells. *Biochimie.* **125**, 223–231.

51. Franceschin, M., Rizzo, A., Casagrande, V., Salvati, E., Alvino, A., Altieri, A., Ciammaichella, A., Iachettini, S., Leonetti, C., Ortaggi, G., Porru, M., Bianco, A., and Biroccio, A. (2012). Aromatic core extension in the series of N-cyclic bay-substituted perylene G-quadruplex ligands: Increased telomere damage, antitumor activity, and strong selectivity for neoplastic over healthy cells. *ChemMedChem.* **7**, 2144–2154.

52. Scaglioni, L., Mondelli, R., Artali, R., Sirtori, F.R., and Mazzini, S. (2016). Nemorubicin and doxorubicin bind the G-quadruplex sequences of the human telomeres and of the c-MYC promoter element Pu22. *Biochim. Biophys. Acta. BBA.* **1860**, 1129–1138.

53. Antonarakis, E.S., and Emadi, A. (2010). Ruthenium-based chemotherapeutics: are they ready for prime time? *Cancer Chemother. Pharmacol.* **66**, 1–9.

54. Alessio, E., Mestroni, G., Bergamo, A., and Sava, G. (2004). Ruthenium antimetastatic agents. *Curr. Top. Med. Chem.* **4**, 1525–1535.

55. Hartinger, C.G., Zorbas-Seifried, S., Jakupec, M.A., Kynast, B., Zorbas, H., and Keppler, B.K. (2006). From bench to bedside — preclinical and early clinical development of the anticancer agent indazolium trans-[tetrachlorobis(1H-indazole)ruthenate(III)] (KP1019 or FFC14A). *J. Inorg. Biochem.* **100**, 891–904.

56. Meng, X., Leyva, M.L., Jenny, M., Gross, I., Benosman, S., Fricker, B., Harlepp, S., Hébraud, P., Boos, A., Wlosik, P., Bischoff, P., Sirlin, C.,

Pfeffer, M., Loeffler, J.-P., and Gaiddon, C. (2009). A ruthenium-containing organometallic compound reduces tumor growth through induction of the endoplasmic reticulum stress gene CHOP. *Cancer Res.* **69**, 5458–5466.

57. He, L., Chen, X., Meng, Z., Wang, J., Tian, K., Li, T., and Shao, F. (2016). Octahedral ruthenium complexes selectively stabilize G-quadruplexes. *Chem. Commun.* **52**, 8095–8098.

58. Grand, C.L., Han, H., Muñoz, R.M., Weitman, S., Von Hoff, D.D., Hurley, L.H., and Bearss, D.J. (2002). The cationic porphyrin TMPyP4 down-regulates c-MYC and human telomerase reverse transcriptase expression and inhibits tumor growth *in vivo*. *Mol. Cancer Ther.* **1**, 565–573.

59. Wu, Q., Zheng, K., Liao, S., Ding, Y., Li, Y., and Mei, W. (2016). Arene Ruthenium(II) complexes as low-toxicity inhibitor against the proliferation, migration, and invasion of MDA-MB-231 Cells through Binding and Stabilizing *c-myc* G-Quadruplex DNA. *Organometallics.* **35**, 317–326.

60. Wang, X., Pei, L., Fan, X., and Shi, S. (2016). [Ru(L)2(3-tppp)]2+ (L=bpy, phen) stabilizes two different forms of the human telomeric G-quadruplex DNA. *Inorg. Chem. Commun.* **72**, 7–12.

61. Betzer, J.-F., Nuter, F., Chtchigrovsky, M., Hamon, F., Kellermann, G., Ali, S., Calméjane, M.-A., Roque, S., Poupon, J., Cresteil, T., Teulade-Fichou, M.-P., Marinetti, A., and Bombard, S. (2016). Linking of antitumor trans NHC-Pt(II) complexes to G-quadruplex DNA ligand for telomeric targeting. *Bioconjug. Chem.* **27**, 1456–1470.

62. Palm, W., and de Lange, T. (2008). How shelterin protects mammalian telomeres. *Annu. Rev. Genet.* **42**, 301–334.

63. Gama, S., Rodrigues, I., Mendes, F., Santos, I.C., Gabano, E., Klejevskaja, B., Gonzalez-Garcia, J., Ravera, M., Vilar, R., and Paulo, A. (2016). Anthracene-terpyridine metal complexes as new G-quadruplex DNA binders. *J. Inorg. Biochem.* **160**, 275–286.

64. Duskova, K., Sierra, S., Arias-Pérez, M.-S., and Gude, L. (2016). Human telomeric G-quadruplex DNA interactions of N-phenanthroline glycosylamine copper(II) complexes. *Bioorg. Med. Chem.* **24**, 33–41.

65. Housaindokht, M.R., and Verdian-Doghaei, A. (2016). Biophysical probing of the binding properties of a Cu(II) complex with G-quadruplex DNA: An experimental and computational study: Probing the binding properties of a Cu(II) complex to G-quadruplex DNA. *Luminescence.* **31**, 22–29.

66. Lauria, A., Bonsignore, R., Terenzi, A., Spinello, A., Giannici, F., Longo, A., Almerico, A.M., and Barone, G. (2014). Nickel(II), copper(II) and zinc(II) metallo-intercalators: Structural details of the DNA-binding by a combined

experimental and computational investigation. *Dalton Trans.* **43**, 6108–6119.

67. Terenzi, A., Bonsignore, R., Spinello, A., Gentile, C., Martorana, A., Ducani, C., Högberg, B., Almerico, A.M., Lauria, A., and Barone, G. (2014). Selective G-quadruplex stabilizers: Schiff-base metal complexes with anticancer activity. *RSC Adv.* **4**, 33245–33256.

68. Biancardi, A., Burgalassi, A., Terenzi, A., Spinello, A., Barone, G., Biver, T., and Mennucci, B. (2014). A theoretical and experimental investigation of the spectroscopic properties of a DNA-intercalator salphen-type ZnII complex. *Chem. Eur. J.* **20**, 7439–7447.

69. Barone, G., Gambino, N., Ruggirello, A., Silvestri, A., Terenzi, A., and Turco Liveri, V. (2009). Spectroscopic study of the interaction of Ni-II-5-triethyl ammonium methyl salicylidene ortho-phenylendiiminate with native DNA. *J. Inorg. Biochem.* **103**, 731–737.

70. Terenzi, A., Ducani, C., Male, L., Barone, G., and Hannon, M.J. (2013). DNA interaction of CuII, NiII and ZnII functionalized salphen complexes: Studies by linear dichroism, gel electrophoresis and PCR. *Dalton Trans.* **42**, 11220–11226.

71. Bonsignore, R., Terenzi, A., Spinello, A., Martorana, A., Lauria, A., Almerico, A.M., Keppler, B.K., and Barone, G. (2016). G-quadruplex vs. duplex-DNA binding of nickel(II) and zinc(II) Schiff base complexes. *J. Inorg. Biochem.* **161**, 115–121.

72. Terenzi, A., Lötsch, D., van Schoonhoven, S., Roller, A., Kowol, C.R., Berger, W., Keppler, B.K., and Barone, G. (2016). Another step toward DNA selective targeting: Ni (II) and Cu (II) complexes of a Schiff base ligand able to bind gene promoter G-quadruplexes. *Dalton Trans.* **45**, 7758–7767.

73. Fleming, A.M., Ding, Y., Alenko, A., and Burrows, C.J. (2016). Zika virus genomic rna possesses conserved G-quadruplexes characteristic of the flaviviridae family. *ACS Infect. Dis.* **2**, 674–681.

Chapter 6

DNA-aided Super-resolution Bioimaging

JiaJun Li*,‡, Giuseppe Arrabito† and ZhaoShuai Gao*

*Shanghai Institute of Applied Sciences
Chinese Academy of Sciences, Jiading Qu,
Shanghai Shi, China
†Department of Physics and Chemistry
University of Palermo, Viale delle Scienze
Parco d'Orleans II, 90128 Palermo, Italy
‡jiajunli@sinap.ac.cn

This chapter reports on the truly-exciting applications of DNA nanoscience in superresolution optical microscopies, showing the ability provided by DNA nanostructures in solving the problems related to the technical difficulties in implementation and ability to achieve higher specificity and multiplexing.

6.1 Introduction

One of the key challenges in bioanalytical sciences is to identify a large number of different molecular species with high temporal and spatial resolution. DNA nanotechnology is able to provide highly sophisticated molecular tools that can facilitate imaging of biological systems with unprecedented resolution and specificity. By employing DNA nanotechnology, one can design fluorescent barcodes with unprecedented spatial control over the fluorophore feature spacing, permitting to produce multiplexing in barcodes, thanks to intensity-coding or geometrical coding. For example, Lin *et al.* [1] reported on the fabrication of DNA-origami based fluorescent barcodes with up to dual-labelled zones to generate six pseudocolors (B, R, G, BG, BR and GR) from three different fluorophores, thus generating $6^3 = 216$ different barcodes.

Biological systems exhibit complex structures at the nanoscale. Understanding the spatial arrangement of their individual components is critical for unravelling the molecular mechanism that underlies complex molecular behavior. Biological specimens can be imaged using Atomic Force Microscopy (AFM) or electron microscopy with a typical resolution in the nanometer and even sub-nanometer regime. However, these methods lack, so far, the ability of non-destructive monitoring of biologically relevant dynamic processes and the determination of kinetics in the sub-second range which can easily be accessed by fluorescence spectroscopy. Although high-speed AFM is able to perform imaging at high speeds [2], it is still limited to a small observation area and is in general much more invasive compared to fluorescence microscopy methods. In principle, while the implementation of optical microscopy approaches with high temporal resolution is straightforward, the possibility to record data with high spatial resolution and over long periods of time due to the diffraction limit and photobleaching has constituted a big challenge for nanometer-scale imaging.

The breakthrough in super-resolution fluorescence microscopy permits today imaging and analyzing biological systems below the classical diffraction limit by switching molecules between fluorescence on- and off-states to obtain sub-diffraction image resolution. In fact, super-resolution fluorescence techniques have bypassed the traditional diffraction limit and demonstrated imaging resolution down to 10–20 nm [3–11]. By pushing optical resolutions down to the molecular scale, super-resolution optical microscopy techniques has permitted to revolutionize biomolecular imaging in cells [12]. Super-resolved images obtained from single molecule detections are in fact the collection of the super-localizations of single emitters imaged successively at low density on the specimen of interest. Here we briefly report on super-resolution fluorescence techniques and the innovations brought in this research field by the application of molecular tools derived from DNA nanotechnology.

6.2 Diffraction Limit

In microscopy, resolution is one of the most important features. In common sense, it means how small an object can be "seen" by a microscope. In a very basic light microscope, the magnified virtual images are created by the combination of an objective lens and an eyepiece lens. By changing the objective lens, the object can be magnified to different levels. Now someone may ask if the object is too small to be observed why not use a bigger magnification objective lens? In a perfect world, lens can focus a beam of light to an infinitely small spot. In that case the intensity of the light in that spot is going be infinitely high and it can destroy everything. You never see any magnifier can do such thing like that don't you? The reason for that is the existence of the optical diffraction limit.

The optical diffraction limit was first found by Ernst Abbe in 1873 [13]. It determines the size of the focal spot (as known as the Point Spread Function, i.e. PSF) when light with certain wavelength passes through an optical lens. It can be described as in the equation below.

$$d = \frac{\lambda}{\sin\theta}$$

In this equation, d is the radius of the PSF, λ is the wavelength of light and $\sin\theta$ is the numerical aperture (NA) of the objective lens. Both the wavelength and NA have their limitations. In fluorescence microscope the signal wavelength is the emission spectrum of the fluorescence molecules and it is usually located in the visible spectrum (450–700 nm). The highest NA objective that can be purchased is featured with a NA of about 1.4. Then it is not hard to find the resolution limit for a conventional optical microscope is at around 200 nm. This resolution is good enough for most of the cell structure imaging but for cell organelles imaging [14] or DNA nanostructures imaging it can hardly qualified for the jobs. There are several techniques tried to break the law of diffraction limit, such as near field microscopy [15], 4pi microscopy [16] and even X-ray microscopy [17], but none of them really succeeded. In these techniques, they either enlarged the NA or reduced the wavelength to push their resolution beyond the 200 nm barrier but they still work under the diffraction limit. As a matter of fact until now scientists are still not be able to truly break this law. In the last few decades, there are three super resolution microscope techniques did successfully "avoided" the diffraction limit and pushed the resolution of the optical microscopy far beyond the 200 nm barrier. Now all of them are commercially produced by the microscope companies and widely used in the bio-imaging area. We are going to introduce the three major super resolution microscopes in the sections below.

6.3 Stimulated Emission Depleting Microscope

The first super resolution microscope we are going to introduce in this book is the stimulated emission depleting (STED) microscope (Figure 6.1). STED microscope is the first super resolution microscope that actually pushed the resolution beyond the optical diffraction limit. Before we talk about the principle of the STED microscope, let us first recall the resolution of a laser-scanning microscope (LSM). The principle of the LSM is to use a highly focused laser beam raster scan across the sample pixel by pixel in x, y directions by using a scanning stage. The laser wavelength is carefully chosen to fit the fluorescence excitation spectrum so it can excite the fluorescent molecules to emit a red shifted fluorescent signal. A single pixel photon sensor such as photon multiplier tube (PMT) or avalanche photodiode (APD) is used as a detector to sense that signal in current mode, voltage mode or photon counting mode to create a contrast image. It is not hard to imagine a picture filled with smaller pixels is going to present more details of the actual sample. As described in the previous section, the size of the PSF is limited by the optical diffraction limit, which is about 200 nm, meaning that any object smaller than 200 nm is going to looks like a 200 nm Gaussian spot in the optical

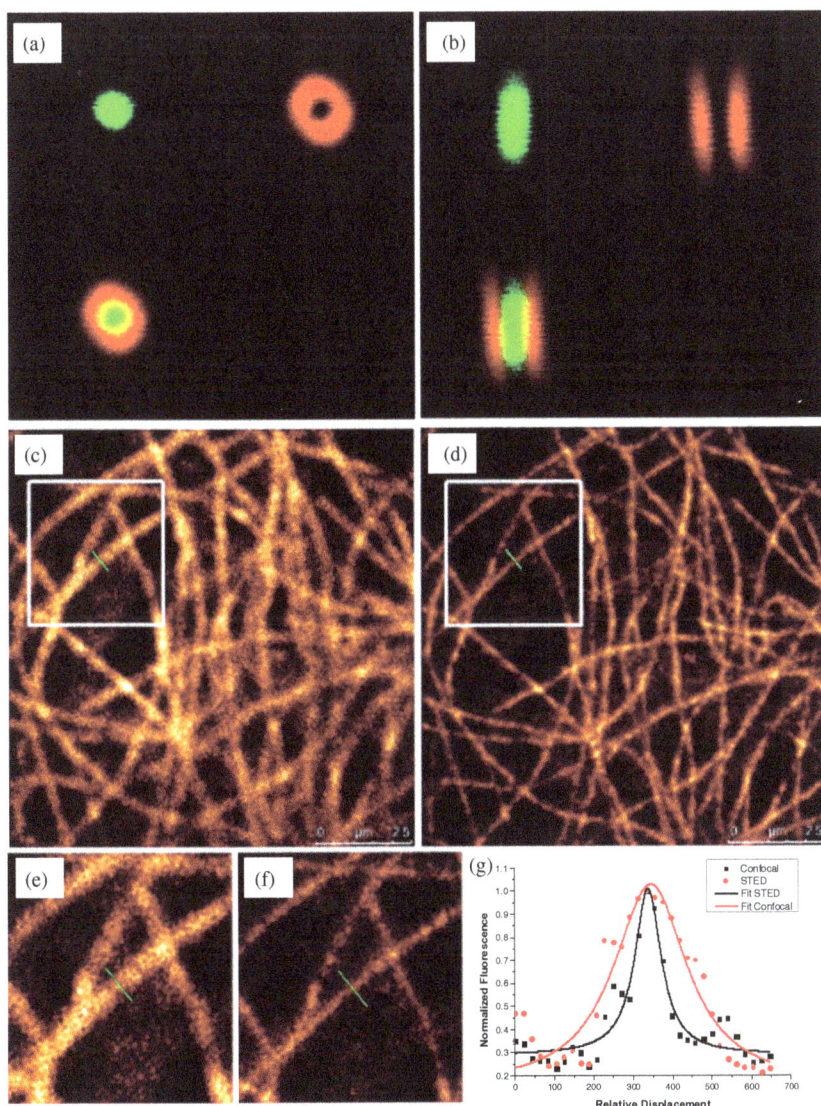

Figure 6.1. The images above are made from a partially customized STED microscope. The donut shape STED depletion beam PSF (red) overlapped on the excitation beam PSF (green) in lateral (a) and longitudinal, (b) directions which enhance the resolution of a confocal image, (c) beyond the diffraction limit (d). (e) and (f) are the zoom of the white squares in (c) and (d). (g) is the intensity profile of a single fibril (alone green lines in (e) and (f)).

microscope. Is there any chance to surpass this limit and confine the PSF beyond the 200 nm? The answer is yes and no. In 1994, Prof. Stefan Hell from the Max Planck Institute for Biophysical Chemistry in Göttingen, Germany brought up an idea to use a depletion beam to compress the efficient excitation PSF to a sub-diffraction limit size. In 1999, he made the first STED microscope and successfully increased the resolution of the optical microscope to 105 nm [18], which was the first time optical microscope resolution went beyond the diffraction limit barrier. Because of this achievement, Prof. Stefan Hell was awarded a Nobel Prize in Chemistry in 2014.

The secret behind the STED microscope is an optical phenomenon called stimulated emission. It happens when a depletion laser PSF spatially and temporally overlapped on an excitation beam PSF. The wavelength of the depletion laser was carefully chosen to be at the end of the fluorescence molecule's emission spectrum. The stimulated emission can compete with the fluorescence emission. If the depletion beam intensity is high enough it is possible to totally eliminate the fluorescence emission. If this happens we say the fluorescence molecule is "switched off" by the depletion beam. Besides of the intensity and the wavelength another important property of the depletion beam is its shape. A classic Gaussian shaped depletion beam can only increase the resolution of the image in one dimension by overlap it on the side of the excitation beam PSF. In the very early stage, Prof. Hell suggested using four depletion beams to squeeze the efficient excitation PSF in a very small point, this plan dramatically increased the complexity of the microscope and it was never been realized. Until later, vortex phase plate was invented, this phase plate can introduce a 0–2π phase delay into the laser wavefront and eventually cause the PSF at the focal point become a ring shape or donuts shape which has a zero-intensity region in the center. Now once this PSF overlapped on the excitation PSF, it "switches off" all the fluorescence molecules except the ones in the zero-intensity center [19]. By overlapping this donut shape depletion beam onto a confocal microscope excitation laser, it can upgrade the confocal microscope to a STED microscope.

Therefore the resolution of the STED microscope is now no longer the size of the excitation PSF, it is depending on the size of the depletion beam's zero-intensity center [20]. The new resolution equation can be written as

$$d = \frac{\lambda}{n\sin\alpha \sqrt{1 + \dfrac{I}{I_{sat}}}}$$

In this equation, λ is the wavelength of the fluorescence signal, $n \sin\alpha$ is the NA of the objective lens, I is the intensity of the depletion beam and I_{sat} is the saturation intensity of the florescence molecule. From the equation we can see that it is very similar to the classic diffraction limit equation except the terms of I and I_{sat}. In theory, if the depletion laser intensity can reach infinitely high, the resolution of the STED microscope is unlimited. In practice it is impossible, not only because there is no infinitely high intensity laser, but also when the laser intensity reaches certain level, it can cause some serious damages to the sample [21, 22]. Therefore it is hard to define an exact number for STED microscope resolution. Usually commercial STED microscope claims their system can reach a resolution of 40 nm, which will depends on what kind of sample you are imaging. The other term in the STED resolution equation is the saturation intensity I_{sat}. It is an instinct property of fluorescence molecules, which is very hard to change, unless to leave the fluorescence molecule in an extremely low temperature. A recent research has been studies that in liquid helium (about −269°C) the resolution of the STED microscope can reach sub 10 nm [23].

Depending on the laser source used for depletion, the STED microscope can be further classified into pulsed laser STED microscope or continue wave laser STED (CW-STED) microscope. The pulsed laser STED microscope is the original design of the STED microscope. It has the benefits of high resolution and low sample damage. In 2007, the CW-STED microscope was invented [24]. The propose of this design is to simplify the complexity of the STED microscope setup by avoiding pulse temporal synchronization. Since CW-STED requires a much higher average intensity depletion laser, this design didn't get widely used. In 2008, a supercontinuum laser was used for excitation source and depletion source, the design gave STED microscope the capability to easily pick the suitable excitation/depletion wavelength pair [25]. Time gating STED (gSTED) microscopes is another member of the STED microscope family. It was first time invented in 2011 [26]. In gSTED microscope, a pulsed laser is used as excitation source and a CW laser is used as depletion source, so there is no need for temporal synchronization. A detector with time gating function was used to remove the fluorescent signal from the first few nanoseconds after the excitation pulses arrive. These signal carries non-superresoltuion information which can reduce the image quality. Due to this design, the STED setup was simplified and its resolution become comparable to the pulsed laser STED microscope. This design now is used in Leica commercial STED microscopes.

6.4 Localization Microscopes

Localization microscopes are another group of super resolution microscopes widely used in bio-imaging area, which include stochastic optical reconstruction microscope (STORM) and photoactivated localization microscope (PALM). They both share a very similar principle. From hardware point of view, the two microscopes are totally identical, the only difference lies in the fact that the PALM was described using photoswitchable fluorescent proteins [6], whereas the STORM was originally described suing Cy5 and Cy3 dyes attached to nucleic acid or proteins. In principle any photoswitchable fluorophore can be used, and the STORM has been demonstrated with a variety of different probes and labeling strategies. The STORM microscope was invented by Prof. Xiaowei Zhuang from Harvard University in United States [27]. The principle of this super resolution technique is totally different to the STED microscope we described in the previous section. If on one hand, the STED microscope is a super resolution microscope modified from a confocal laser scanning microscope by modifying its hardware, the STORM microscope is a super resolution microscope modified from a total internal reflection fluorescence (TIRF) microscope by adding some specific algorithms in it (Figure 6.2). TIRF microscope is a very commonly used fluorescence microscope. The difference between TIRF and a wide field fluorescence microscope is the way it illuminates the sample. A wide field fluorescence microscope illuminates a sample by sending the excitation beam directly onto it, by doing

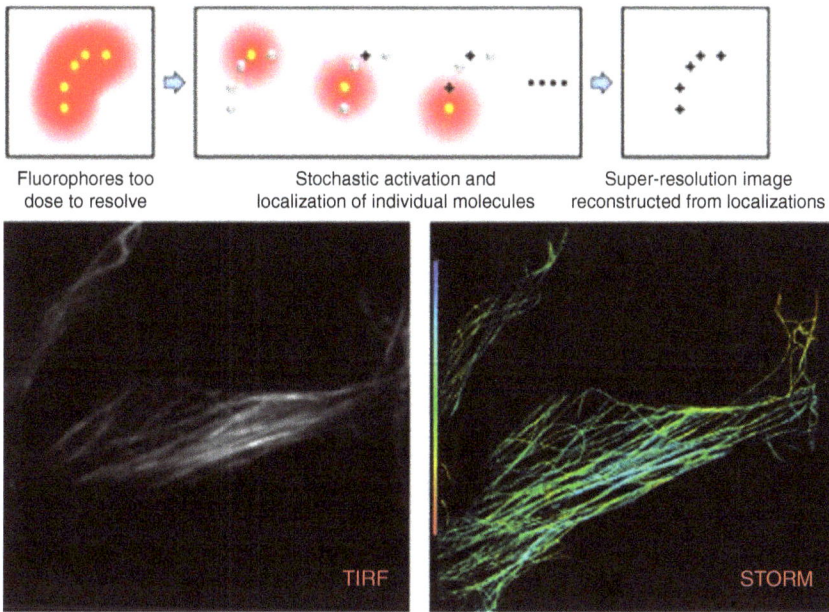

Figure 6.2. The principle of the STORM microscope is to calculate the localization of every fluorescent molecules (upper). After the image reconstruction, the resolution is dramatically enhanced compare the to the conventional TIRF microscope (lower).

this the fluorescent molecules emit the florescent signal and these signals are received by a CCD or CMOS camera. A disadvantage of the wide field fluorescence microscope is during the imaging process the entire sample is illuminated by the excitation source. The fluorescent molecules from different layers of the sample produce huge amount of signal and the detector receives these signals spontaneously. The microscope couldn't distinguish which layer does the signal come from. This significantly decreases the signal to noise ratio, especially in the thick tissue imaging. In TIRF microscope the illumination source is the evanescent wave of total internal reflection (TIR). The TIR happens when the laser incident angle went beyond the critical angle (Snall's law). This evanescent wave is only few hundreds thick and only illuminates a very thin layer of sample. This technique reduces the image background noise and enables single molecule fluorescence imaging. The capability of the single fluorescence molecule imaging is the key to the STORM super resolution microscopy.

As described in previous sections, no matter how small the objects are, if the sizes of them are smaller than 200 nm, in an optical system they only can be seen as a 200 nm Gaussian spot. Though the size of the PSF cannot be changed, it is still possible to localize its center position. The standard error of the localization can be described in the equation below [28].

$$\sigma = \sqrt{\left(\frac{s^2}{N}\right) + \left(\frac{a^2/12}{N}\right) + \left(\frac{8\pi s^4 b^2}{a^2 N^2}\right)}$$

In this equation, s is the standard deviation of the PSF Gaussian function, N is the number of photons captured from the single fluorescent molecule, a is the pixel size of the detector and b is the standard deviation of the background. Once the photon number reaches 1000, the equation can be approximated as

$$\sigma \approx s/\sqrt{N}$$

Therefore now the standard error is only related to the standard deviation of the PSF Gaussian function and the number of photons captured from the single fluorescent molecule. In an ideal world, if a fluorescent molecule stays absolutely immobilized and can produce infinite number of photons, it can be localized to an infinitely small spot, sometimes even smaller than the size of the single molecule itself. Even in practice it is still possible to keep the localization distribution in a few nanometer scale, which is much smaller than the diffraction limit. This method is ideal for single fluorescent molecule tracking but still not good enough for super resolution imaging. In fluorescence imaging the fluorescent molecules are distributed very close. If the distance between two fluorescent molecules is smaller 200 nm and they are excited spontaneously, they cannot be resolved in a single frame, which makes single molecule localization impossible. Therefore another key component in the STORM microscope is to make fluorescent molecules "blink," so they can be resolved temporally. The situation can be described as in a paintball game if one shoots all the bullets close to the center of a bullseye, eventually he will get a paper full with paint spots. Since those paint spots are so close and overlapped on each other you are not able tell the localizations of each bullet. If at the meantime, one used a high speed camera captured the whole shooting process, it is possible to stop at each frame when a single bullet splashed on the paper and calculate the localization of its "PSF." By analyzing every frame it is possible to draw all the bullets localizations on a new paper. This is the secret behind the STORM microscope.

The trick is to control the fluorescent molecules "blink" rate to make sure there is a very low possibility two close fluorescent molecules be excited spontaneously in a single frame. There are a lot of commercial fluorescent dyes on the market which are designed for this purpose [29], which covers almost all the common spectrums.

In order to obtain a decent super resolution image usually 30,000–50,000 frames are needed. Just like STED microscope, its super resolution efficiency is highly dependent on the sample and the fluorescent dye. In most of the cases the resolution can reach up to 20 nm [30, 31], which is slightly better than the STED microscope. One has to consider that the imaging time is extremely long, which makes the STORM impossible to do *in vivo* imaging. Since the STORM microscope's illumination technique is based on TIRF illumination, it was considered lack of the 3D imaging capability. In recent years, some microscope company use a cylindrical lens with 3D algorithm realized the STORM whole cell imaging.

6.5 Structured Illumination Microscope

The last super resolution microscopy that we here describe is the SIM microscope. The Structured Illumination Microscope (SIM) microscope was invented by Prof. M. G. L. Gustafsson in 2000 [29]. The principle of SIM microscope is the combination of structured illumination and Fourier transform [32]. With respect to most of the microscopes, the SIM's illumination source is a laser with a sinusoidally striped illumination patterns. When this pattern overlaps on an unknown pattern e.g. the structure of the fluorescent molecules distribution from the sample, it produces moiré

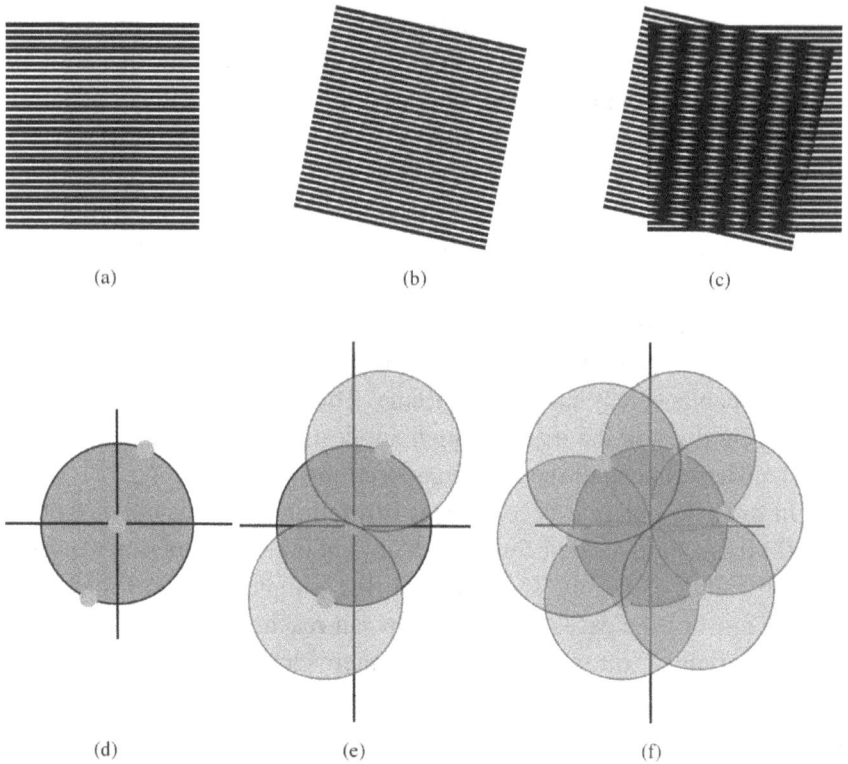

Figure 6.3. When two structured illumination overlap on each other in certain angle (one for the sample (a) another one for the structured illumination source (b)), they result a moiré fringes (c). An extremely fine sinusoidally pattern (200 nm between each line) creates two bright spots and an origin in its Fourier transform image. The distance between the two bright spots determines the size of the observation region (a). The moiré fringes provides information outside of the observation region (b). By rotating the sinusoidal illumination several time radius of the observation region of the optical microscope can be doubled therefore it increases the microscope resolution.

fringes. Moiré fringes are enlarged scale interference patterns, which usually happen when two unidentical patterns overlap on each other as shown in Figure 6.3(a). This enlarged pattern contains information of both the illumination pattern and the sample pattern.

Under the diffraction limit, if sample structure is smaller than 200 nm, it cannot be resolved by the optical microscope, but when they multiplied with the structured illumination source the moiré fringes may become observable. Since one already knows the structure the illumination source and the moiré fringes, we can extract the structure of the sample. Different to the previous two super resolution microscopes, the resolution of the SIM microscope is settled. In image processing, all images can be transformed to its Fourier transform, a mathematical function that transforms image from its spatial domain to its frequency domain. In frequency domain the high frequency components (the components away from the origin) determines image's "detail" and the low frequency components (the components close to the origin) determines image's "gray scale." If one has a sinusoidally pattern as image, in its Fourier transform image it is presented as two mirrored bright spots offset away from the origin. The directions of the two spots are respect to the directions of the strips of the pattern and the distance between the spots to the origin are proportional to the inverse line spacing of the pattern. If the lines spacing are exactly at the diffraction limit and rotating extremely fast, in its Fourier transform image it is presented as a circle constructed by the two bright spots. The components within this circle are called the "observation region" within the optical diffraction limit. Since the higher frequency components determine image's detail, all the components outside of this circle are "super resolution region." In SIM microscope the moiré fringes provide information outside of the observation region. In its Fourier transform, two extra bright spots are centered by the two original bright spots. Since the two original bright spots are already at the edge of the observation region, by the rotating the direction of the sinusoidally striped illumination several times (usually 3 times) the SIM microscope is able to double the resolution of the traditional optical microscope in x, y directions.

6.6 Structured Overview about Super Resolution Technologies

The three super resolution technologies described in the previous sections are the most widely used super resolution microscopes. All of them are commercially produced and have been served in the bio imaging area for the last a few years. There is no simple way to say which microscope is the best. All of them have their own pros and cons. It is important to choose the right microscope, which ideally fits samples condition [33]. Here is a table may give users some ideas to the most suitable microscope.

	STED	**STORM**	**SIM**
Resolution	60 nm	40 nm	100 nm
Speed	Fast	Slow	Very fast
***In vivo* imaging**	Possible	Impossible	Possible
Damage to the sample	High	Low	Low
3D imaging	Possible	Possible	Impossible
3D Super resolution	Possible	Possible	Impossible
Multi-color imaging	Conditions may applied	Possible	Possible

6.7 DNA-nanostructures Aided Super Resolution Imaging

There are many parameters which affect the performance of current super-resolution techniques in imaging complex biological systems. Just to mention the most important factors, one can consider the fluorophore photon budget, suboptimal fluorophore imaging efficiency and the limited control over target blinking kinetics. Such restrictions bring to limited photon count per localization, limited number of blinking events per target and high fraction of false localizations which finally lead to limited number of visualized blinking events per target, affecting signal-to-noise ratio and the visualization of individual targets within dense clusters. Methods such as PALM or STORM have difficulties to access very densely labeled regions due to spontaneous photoswitching, which prevents imaging individual fluorophores in these regions [34].

In the following, we describe two super resolution microscopy techniques in which DNA nanostructures play a fundamental role. In particular, we will describe Binding-activated localization microscopy (BALM) and Points accumulation for imaging in nanoscale topography (PAINT).

6.7.1 *Binding-activated localization microscopy*

BALM [35], is a superresolution technique in which free solution diffusing dyes turn bright upon binding to a target structure and can be localized with

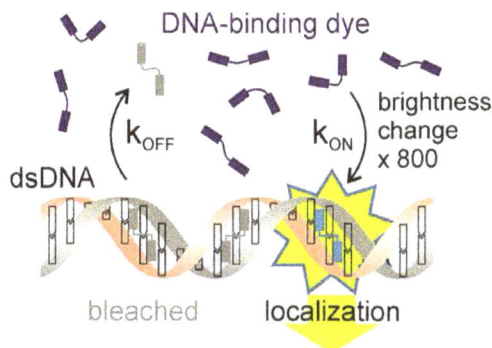

Figure 6.4. Scheme of BALM. Dyes diffusing in solution phase turn bright when they bind to DNA target structures. Binding kinetics is determined by the on rate k_{on}, whereas detaching is determined by k_{off}. Reprinted with permission from Ref. [35]. Copyright (2011) American Chemical Society.

sub-diffraction resolution before they eventually bleach or detach. Upon binding to target object, the chromophores become immobilized and the quantum yield increases of 2–3 orders of magnitude (Figure 6.4).

In their paper [35] Schoen *et al.* improved upon both localization accuracy and density by utilizing fluorophores that are "switched on" upon binding to a target structure and directly exploit this property to localize them under dynamic binding conditions (Figure 6.5). By iterating cycles of binding, localization and bleaching, it is possible to reconstruct the biological structure under investigation. Such strategy is defined as BALM and can be generalized to visualize biological or synthetic structures with nanoscale resolution using a variety of available dyes. By the employment of a dynamic labeling scheme and the optimization of fluorophore brightness, they obtained a resolution of ~14 nm (fwhm) and a spatial sampling of 1/nm.

The authors used YOYO-1 dye to image double-strand DNA with unprecedented high localization accuracy and high labeling density at the same time. Subsequently, they used the dye PicoGreen to image chromosomal organization in fixed *Escherichia coli* cells, obtaining a dramatic improvement in resolution when compared with diffraction-limited microscopy. The chromosome in such cells and the overall contour of the nucleoid inside the cells obtaining details with a resolution on the order of about 100 nm.

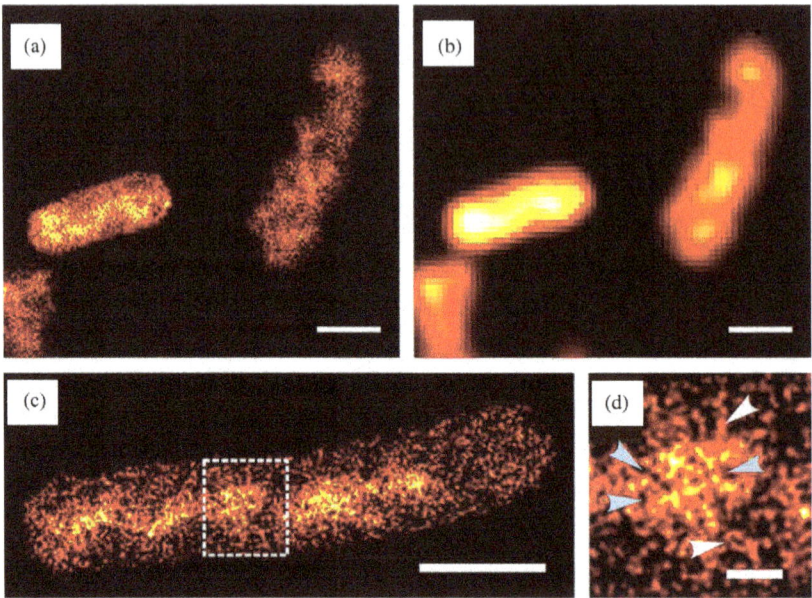

Figure 6.5. Imaging of chromosomes in fixed bacteria by BALM with PicoGreen fluorophore. (a) Image obtained via BALM, (b) diffraction-limited image, (c) image of bacteria during division, (d) enlarged view. Scale bars: 1 μm (a)–(c), 200 nm (d). Adapted with permission from Ref. [35]. Copyright (2011) American Chemical Society.

6.7.2 *Binding-DNA-aided point accumulation for imaging in nanoscale topography*

6.7.2.1 *Points accumulation for imaging in nanoscale topography*

In the PAINT method [11], the object to be imaged has to be continuously targeted by fluorescent probes which are present in the solution phase. Notably, fluorescent probes flux on the object is dependent on molecular diffusion coefficient and the concentration gradient of the probes. A fluorescent signal on the object to be imaged is constituted by a diffraction-limited spot. Such signal can be destroyed if the fluorescent probe is photobleached or if it dissociates from the object. In this original version of PAINT, the probe does not necessarily have a specific interaction with the object, this meaning that such interaction can be electrostatic coupling, hydrophobic interaction. The authors also showed that the collision rate of probes with the object depends linearly on the concentration of the probes. Such fluorescence spikes appearance can be approximated by the diffusion-controlled bimolecular reaction rate constant multiplied to the probability that a collision is able to generate a binding configuration leading to a fluorescent signal. As one can expect, the nature of local discrimination in imaging is dependent on the specific interaction between fluorescence probe and the object.

In PAINT, it is necessary to keep density of probes on the object surface below the value of one molecule per squared micron in each observation period. For this reason, the obtained fluorescent image is treated as a PSF from a single molecule. In order to achieve this condition, the time between subsequent observation frames has to be longer than the time of fluorescence burst of probe and by maintaining the image frame length shorter than the period between collisions.

Probing molecules are deactivated between successive image recordings. The authors demonstrated the impressive 25 nm resolution on lipid bilayers, contours of such bilayers and contours of such lipid bilayers and large unilamellar vesicles. In particular, they used a fluorescently labeled transferrin protein and a small environment-sensitive dye (Nile red) to demonstrate the possibility to use different probes by employing this method. Remarkably, PAINT can be used without any modification of the imaging objects such as cellular compartments or organelles or cellular membranes. By varying the probe molecule, it is possible to obtain selectivity in imaging specific objects. For instance, many fluorescently labeled proteins which have high specificity to certain cellular compartments can be used as imaging probes.

In other words, the number of molecules targeting a circular area πR^2 in unit time from a solution which contains N molecules per unit volume is given by a Smoluchowski type relation, $4DRN$ (s^{-1}), in which D stands for molecular diffusion coefficient. Therefore, in the steady state, the number of probes on the object is given by $4DRN <\tau_{on}>$, in which $<\tau_{on}>$ is the mean time that the probe spends with the object before it dissociates from it or it is photobleached. The deactivation rate $<\tau_{on}>^{-1}$ is defined as the sum of the mean rates of photobleaching and dissociation of the probe from the object, by assuming that the probes act independently of one another. Bleaching rate is a function of the excitation intensity and the photochemical and photophysical properties of the probe in the surrounding medium. As expected, it is possible to control the time that the probe spends in contact with the object by varying the concentration of the probe in solution. Since the number of probe molecules in the surrounding solution is not significantly depleted by the photobleaching occurring on only one small illuminated object, it is possible to acquire the image by accumulating as many images frames as needed. The image is formed by overlapping such frames. The total recording time is limited by the mechanical stability and movements of the object to image with respect to the light collection system.

6.7.2.2 *DNA-aided PAINT*

As previously explained in PAINT approach, imaging is carried out by the employment of diffusing fluorescent molecules that interact transiently with the sample. A key limit of PAINT is that interaction of dyes with the sample occurs via electrostatic or hydrophobic coupling, so it cannot be used in multiplexed analysis of different biomolecules in the same sample. In order to achieve programmable multiplexed dye interactions and to increase the specificity and the number of usable fluorophores, DNA-PAINT was developed [36, 37]. In DNA-PAINT, stochastic switching between fluorescence on- and off-states is executed via transient binding of fluorescently labeled oligonucleotides ("imager" strands) to complementary "docking" strands on DNA nanostructures. In the unbound state, it is possible to observe only background fluorescence from imager strands. In case of binding and immobilization of an imager strand, fluorescence emission is detected. In DNA-PAINT, the transient binding of fluorophore-labelled imager strands to target-bound docking strand permits to achieve the necessary blinking condition for superresolution reconstruction. Remarkably, the continuous replenishment of imager strands renders DNA-PAINT immune to photobleaching, allowing high localization precision by extracting a large number of photons per single-molecule localization and a high target

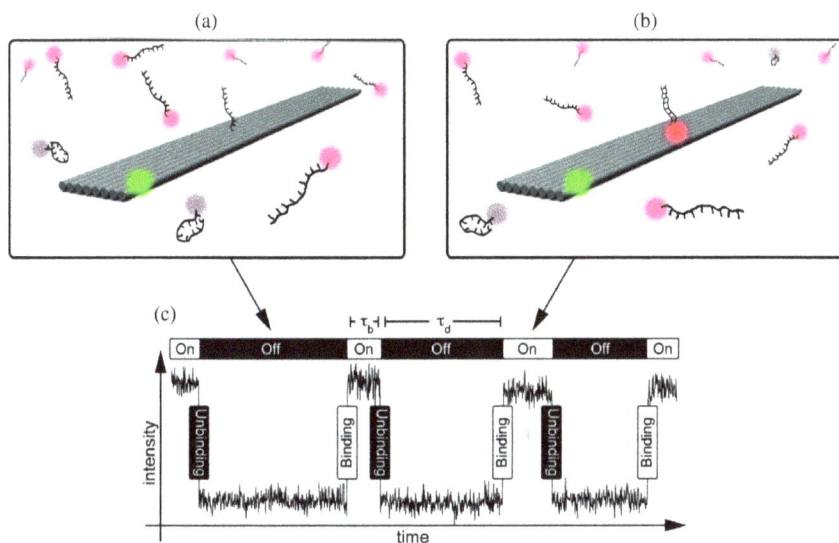

Figure 6.6. Scheme of DNA-origami structures (nominally 291.5 × 23 nm) employed for the single-molecule binding. Each structure contains a fluorescently labeled staple strand (labeled with ATTO532). The docking strand extension is shown at the center of the structure. Without binding of the imager strand, no fluorescence signal is observed. (a) Without binding of the imager strand, no fluorescence signal is observed. (b) Upon hybridization of an imager strand, fluorescence signal was observed in the red channel. (c) Trace of fluorescence intensity vs. time for the unbound state (τ_d is the time for the dissociated state). Adapted with permission from Ref. [36]. Copyright (2010) American Chemical Society.

separability by collecting a large number of blinking events from each target. Additionally, due to independent and programmable control of blinking on- and off- rates, DNA-PAINT permits low imaging background in dense clusters from appropriately adjusted blinking duty cycle based on the target density.

In a fundamental paper, Jungmann *et al.* [36] implemented the PAINT concept by means of reversible labeling with diffusing DNA probes (see Figure 6.6). They defined this approach as DNA-PAINT. This uses fluorescently labeled single stranded DNA probes in solution that hybridize to complementary single-stranded extension on DNA-origami structures, thus imaging them in real time. Single-molecule DNA binding and unbinding events are directly visualized using fluorescence microscopy, making DNA-PAINT, apart from its imaging capabilities, an ideal tool for analyzing, e.g. hybridization kinetics on DNA structures.

They fabricated long rectangular 2D origami structures with dimensions of about 290×23 nm. Each origami structure is decorated with a fluorescently labeled staple strand (ATTO532 labeled) and one or more docking strands. They firstly acquired a diffraction-limited fluorescence image using the directly incorporated fluorescent label (here ATTO532 using an excitation wavelength of 532 nm) to determine the positions of the origami structures on the surface. Then, they added ATTO655 labeled imager strands having a complementary sequence to the docking strands. They selected lengths of the imager strands just to result only in short-term binding events at ambient conditions. A second excitation wavelength (650 nm for ATTO655) is used to monitor the transient binding of single imager strands to the DNA origamis. From the measurements, they could assign low signals corresponding to unoccupied staple strands, whereas binding of an imager strand resulted in fluorescence increase. The perfect superposition of the green marker image (ATTO532) with a standard deviation image that highlights the imager strand (ATTO655) locations based on the maximum signal variations indicates the high selectivity of binding to DNA origami. Given the possibility to monitor individual hybridization events on a DNA origami in real time, they could calculate the association rate from the τ_d, which is defined as the mean interevent lifetime (i.e. the dissociated time) since $1/\tau_d = k_{on}c$, in which c is the imager strand concentration.

They could also determine imaging efficiency, given the precise control over number and position of docking strands on monomeric DNA-origami structures. The authors could determine the fraction of docking strand positions that could be imaged. The imaging efficiency is actually an open challenge for superresolution approaches, for many reasons like not precise labeling in some positions, bleaching of fluorophores during photoactivation or not sufficient number of emitted photons before photobleaching. For example, in PALM, only a percentage of about 30% of molecules is usually analyzed [6]. The authors could determine that almost all docking strands can be imaged by comparing the presence of staple strands at specific positions using AFM and DNA-Paint measurements. In this regard, after having introduced three biotinylated DNA strands in the DNA origami, streptavidin was added to the origami sample on a mica surface and incubated for several hours. They imaged structures containing three or two streptavidin molecules, since only these structures can be counted by DNA-PAINT. Remarkably, the efficiency of imaging was found to be almost complete (around 95%), this value is significantly higher than current super-resolution techniques.

Jungmann *et al.* reported on an unprecedented methodology that carries out the transient binding of short fluorescently labeled oligonucleotides (DNA-PAINT, a variation of point accumulation for imaging in nanoscale topography) allowing for multiplexed super-resolution imaging at sub-10 nm spatial resolution *in vitro* on synthetic DNA structures [38]. They demonstrated superresolution imaging of microtubules inside HeLa cells using Atto 655-labeled imager strands. Remarkably, they defined a multiplexing approach (Exchange-PAINT) since it allows sequential imaging of multiple targets inside single cells using only a single dye and a single laser source. In exchange-PAINT only one fluorophore is employed, thereby allowing the selection of an optimal dye with respect to its photophysical properties for superresolution imaging. They obtained super-resolution images of β-tubulin in microtubules, COX IV in mitochondria, TGN46 in the Golgi complex and PMP70 in peroxisomes using Cy3b dye (see Figure 6.7(a)).

In a very recent article, Dai *et al.* [39] reported on the combination of "discrete molecular imaging" with Exchange-PAINT showing highly accurate (<1 nm) three-color registration, in addition to highly accurate drift correction (<1 nm r.m.s.) within each channel. The necessary condition for "discrete molecular imaging" is the high localization precision that allows a full width at half maximum (FWHM) resolution equal to or smaller than the spacing between them.

They could visualize individual targets in a compactly labeled molecular grid of targets (with a point-to-point spacing of ~5 nm) demonstrating multiplexed imaging on a three-color nanodisplay board with ~5 nm pixels. They showed the potentiality of their method by imaging a custom-designed letter pattern ("Wyss!") on a 60×85 nm origami nanodisplay breadboard with 5 nm display pixel size. Finally, they imaged a three-color mixture structure of the 5 nm grid with multiplexed DMI with an average localization precision of 2.0 nm (see Figure 6.7(b)). Their

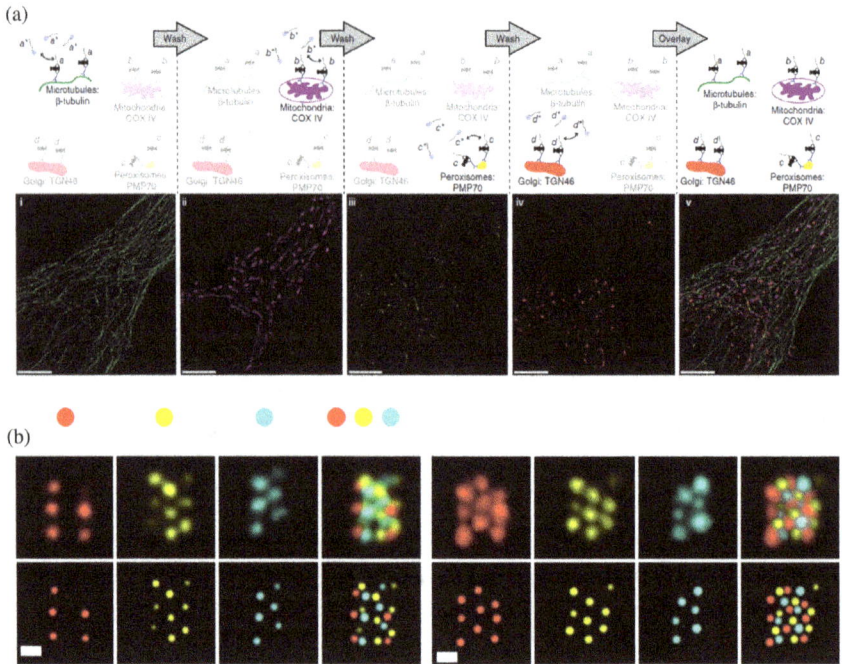

Figure 6.7. (a) Multiplexed cell analysis by DNA-PAINT. Each target in a fixed HeLa is labeled with an antibody carrying a unique DNA-PAINT docking sequence, recognized by a specific Atto 655 labeled imager strand. Super-resolution images of β-tubulin in microtubules (i), COX IV in mitochondria (ii), TGN46 in the Golgi complex (iii) and PMP70 in peroxisomes (iv) were obtained sequentially; (v) overlay of all four targets. Scale bar is 500 nm. Reprinted with permission from Macmillan Publishers Ltd.: Nature Methods (Ref. [38]), copyright © (2014). (b) Discrete molecular imaging with complex patterns and multiplexed (three color) images. DNA-PAINT super-resolution images (top row) and automatically fitted images (bottom row) are shown for all three single-color channels (left three columns) and the combined image (right column) for two representative 5 nm grid structures. Reprinted with permission from Macmillan Publishers Ltd.: Nature Nanotechnology (Ref. [39]), copyright © (2016).

approach can potentially enable the determination of the position and identity of discrete molecular components in complex biological or synthetic nanostructured systems, so allowing for a complementary approach to electron microscopy and crystallography with single-molecule sensitivity. According to the authors, two big issues remain for future development of their methodology. The first one deals with the necessary trade-off between spatial and temporal resolution, i.e. the compromise between blinking times (higher times lead to higher spatial resolution) and image acquisition time (higher times lead to slower acquisitions). The other issue is the limited absolute labelling efficiency (i.e. average number of probes labeled per molecular target), which can be overcome by employing genetically labeled tags (such as SNAP-tags), aptamers, small-molecules labels just to cite the most important ones. The future development of their methodology allows for investigation of molecular features in diverse biological systems such as the molecular composition and architecture of complex cellular systems (such as molecular states of individual protein components or architecture of chromosomes) with unprecedented spatial resolution.

6.8 Conclusion

In Conclusion, DNA nanostructures allow for a significant simplification in currently established superresolution optical microscopies, solving the problems related to the technical difficulties in implementation. Even more importantly, DNA allows for truly multiplexing for a large number of distinct biomolecular targets. From the most recent literature reports, one can expect that in the future, it will be possible to quantitatively study molecular features in diverse crowded environments (cytosol, cellular membranes) allowing for unprecedented analysis of biological systems at nanometer scale resolution.

References

1. Lin, C., Jungmann, R., Leifer, A.M., Li, C., Levner, D., Church, G.M., Shih, W.M., and Yin, P. (2012). Submicrometre geometrically encoded fluorescent barcodes self-assembled from DNA. *Nat. Chem.* **4**, 832–839.
2. Endo, M., Katsuda, Y., Hidaka, K., and Sugiyama, H. (2010). Regulation of DNA methylation using different tensions of double strands constructed in a defined DNA nanostructure. *J. Am. Chem. Soc.* **132**(5), 1592–1597.
3. Hell, S.W., and Wichmann, J. (1994). Breaking the diffraction resolution limit by stimulated emission: Stimulated-emission-depletion fluorescence microscopy. *Opt. Lett.* **19**, 780–782.
4. Klar, T.A., and Hell, S.W. (1999). Fluorescence microscopy with diffraction resolution barrier broken by stimulated emission. *Opt. Lett.* **24**, 954–956.

5. Gustafsson, M.G. (2000). Surpassing the lateral resolution limit by a factor of two using structured illumination microscopy. *J. Microsc.* **198**, 82–87.

6. Gustafsson, M.G. (2005). Nonlinear structured-illumination microscopy: Wide-field fluorescence imaging with theoretically unlimited resolution. *Proc. Natl. Acad. Sci. USA.* **102**, 13081–13086.

7. Betzig, E., *et al.* (2006). Imaging intracellular fluorescent proteins at nanometer resolution. *Science.* **313**, 1642–1645.

8. Hess, S.T., Girirajan, T.P., and Mason, M.D. (2006). Ultra-high resolution imaging by fluorescence photoactivation localization microscopy. *Biophys. J.* **91**, 4258–4272.

9. Rust, M.J., Bates, M., and Zhuang, X. (2006). Sub-diffraction-limit imaging by stochastic optical reconstruction microscopy (STORM). *Nat. Meth.* **3**, 793–795.

10. Heilemann, M., *et al.* (2008). Subdiffraction-resolution fluorescence imaging with conventional fluorescent probes. *Angew. Chem. Int. Ed.* **47**, 6172–6176.

11. Sharonov, A., and Hochstrasser, R.M. (2006). Wide-field subdiffraction imaging by accumulated binding of diffusing probes. *Proc. Natl. Acad. Sci. USA.* **103**, 18911–18916.

12. Szymborska, A., *et al.* (2013). Nuclear pore scaffold structure analyzed by super-resolution microscopy and particle averaging. *Science.* **341**, 655–658.

13. Abbe, E. (1873). Beiträge zur Theorie des Mikroskops und der mikroskopischen Wahrnehmung. *Archiv für mikroskopische Anatomie.* **9**, 413–418.

14. Fernández-Suárez, M., and Ting, A.Y. (2008). Fluorescent probes for super-resolution imaging in living cells. *Nat. Rev. Mol. Cell Biol.* **9**, 929–943.

15. Hecht, B., *et al.* (2000). Scanning near-field optical microscopy with aperture probes: Fundamentals and applications. *J. Chem. Phys.* **112**, 7761–7774.

16. Hao, X., *et al.* (2015). Point-spread function optimization in isoSTED nanoscopy. *Opt. Lett.* **40**, 3627–3630.

17. Rambo, R.P., and Tainer, J.A. (2013). Super-resolution in solution X-ray scattering and its applications to structural systems biology. *Ann. Rev. Biophys.* **42**, 415–441.

18. Klar, T.A., *et al.* (2000). Fluorescence microscopy with diffraction resolution barrier broken by stimulated emission. *Proc. Natl. Acad. Sci. USA.* **97**, 8206–8210.

19. Willig, K.I., *et al.* (2006). Nanoscale resolution in GFP-based microscopy. *Nat. Meth.* **3**, 721–723.

20. Hofmann, M., *et al.* (2005). Breaking the diffraction barrier in fluorescence microscopy at low light intensities by using reversibly photoswitchable proteins. *Proc. Natl. Acad. Sci. USA.* **102**, 17565–17569.

21. Donnert, G., Eggeling, C., and Hell, S.W. (2007). Major signal increase in fluorescence microscopy through dark-state relaxation. *Nat. Meth.* **4**, 81–86.
22. Boudreau, C., *et al.* (2016). Excitation light dose engineering to reduce photo-bleaching and photo-toxicity. *Sci. Rep.* **6**, 30892.
23. Yang, B., *et al.* (2015). Optical nanoscopy with excited state saturation at liquid helium temperatures. *Nat. Photon.* **9**, 658–662.
24. Willig, K.I., *et al.* (2007). STED microscopy with continuous wave beams. *Nat. Meth.* **4**, 915–918.
25. Wildanger, D., *et al.* (2008). STED microscopy with a supercontinuum laser source. *Opt. Express.* **16**(13), 9614–9621.
26. Engelhardt, J., *et al.* (2011). Sharper low-power STED nanoscopy by time gating. *Nat. Meth.* **8**(7), 571–U75.
27. Rust, M.J., Bates, M., and Zhuang, X. (2006). Sub-diffraction-limit imaging by stochastic optical reconstruction microscopy (STORM). *Nat. Meth.* **3**(10), 793–796.
28. Yildiz, A., *et al.* (2003). Myosin V Walks hand-over-hand: Single fluorophore imaging with 1.5-nm localization. *Science.* **300**(5628), 2061.
29. Dempsey, G.T., *et al.* (2011). Evaluation of fluorophores for optimal performance in localization-based super-resolution imaging. *Nat. Meth.* **8**(12), 1027–1036.
30. Bates, M., *et al.* (2007). Multicolor super-resolution imaging with photoswitchable fluorescent probes. *Science.* **317**(5845), 1749–1753.
31. Huang, B., *et al.* (2008). Three-dimensional super-resolution imaging by stochastic optical reconstruction microscopy. *Science.* **319**(5864), 810–813.
32. Rego, E.H., *et al.* (2012). Nonlinear structured-illumination microscopy with a photoswitchable protein reveals cellular structures at 50-nm resolution. *Proc. Nat. Acad. Sci. USA.* **109**(3), E135–E143.
33. Graydon, O. (2016). View from ICON Europe 2016: Next step for super-resolution. *Nat. Photon.* **10**(8), 504–505.
34. Geissbuehler, S., Dellagiacoma, C., and Lasser, T. (2011). *Biomed. Opt. Express.* **2**, 408–420.
35. Schoen, I., Ries, J., Klotzsch, E., Ewers, H., and Vogel, V. (2011). Binding-activated localization microscopy of DNA structures *Nano. Lett.* **11**(9), 4008–4011.
36. Jungmann, R., Steinhauer, C., Scheible, M., Kuzyk, A., Tinnefeld, P., and Simmel, F.C. (2010). Single-molecule kinetics and super-resolution microscopy by fluorescence imaging of transient binding on DNA origami. *Nano Lett.* **10**, 4756–4761.

37. Raab, M., Schmied, J.J., Jusuk, I., Forthmann, C., and Tinnefeld, P. (2014). Fluorescence Microscopy with 6 nm resolution on DNA Origami. *ChemPhysChem.* **15**, 2431–2435.

38. Jungmann, R., Avendano, M.S., Woehrstein, J.B., Dai, M., Shih, W.M., and Yin, P. (2014). Multiplexed 3D cellular super-resolution imaging with DNA-PAINT and Exchange-PAINT. *Nat. Meth.* **11**, 313–318.

39. Dai, M., Jungmann, R., and Yin, P. (2016). Optical imaging of individual biomolecules in densely packed clusters. *Nat. Nanotechnol.* **11**, 798–807.

Conclusion

In Conclusion, we hope to have provided the reader of the book with a comprehensive idea of what DNA can do in many different aspects of nanosciences and life sciences, a part from being the molecule that stores all the genetic information in living cells.

When one thinks about DNA, what comes to mind is the concept of the fundamental molecule that stores the genetic material in all the living systems. In principle, any molecule capable to interfere with the DNA, blocking its function may therefore become lethal for a cell. For such reason, DNA has been widely considered an important target to halt cell proliferation, especially in life-threating diseases as cancer. For instance, non-canonical forms such as G-quadruplex DNA play an important role in the transcription of genes and in the replication of human genome, and in this book we also discuss about these peculiar molecules. What it is known also to non-specialist is that DNA possesses many unique properties, including its precise programmability between complementary strands, fundamental biological function, biocompatibility and manipulation due to a vast toolkit of enzymes for cutting, ligation, polymerization and so on.

These possibilities to assemble programmable molecular nanostructures by using Watson–Crick base pairing — being such interaction predictable, reliable and reversible — permitted the onset of DNA nanotechnology in the early 1980s from the ideas of Nadrian Seeman, a physicist who was attempting to crystallize proteins for *X*-ray diffraction studies and thought of using DNA molecules as a tool for creating an artificial lattice having the possibility to arrange proteins in periodic locations. From those years, DNA has been shown as tool to fabricate ordered nanostructures that can be then functionalized/templated with different biomolecules/nanomaterials as different as nanoparticles, nanowires, organic molecules, peptides, proteins with controlled spacing at the nanometer scale (<10 nm) and with properties depending upon exposition to external stimuli (light, temperature, pH, etc.). In this way, one can combine the properties of both DNA and nanomaterials for exposing designed functionality and customizable geometrical hetero-nanostructures. These smart molecular systems can give rise to unprecedented tools that can revolutionize the analytical arena due to its availability of sensing and detecting multiple analyzes with unprecedented control at nanoscale resolution. One can expect that by coupling automated on-chip high yield DNA synthesis with low cost detection methods, DNA-nanotechnology is potentially able to realize ultra-high-sensitivity, multiplexed bioanalytical assays for many different applications like diagnostics, drug screening, toxicology, immunology and biosensors with ultra-low reagent consumption.

A fundamental aspect for the application is the ability to integrate DNA nanostructures into top-down fabricated electronic devices allowing for the extraction of electrical analog/digital signal as analytical output. Such nanostructures can represent the core part of high-sensitivity biosensor devices with unprecedented analytical performances when compared to those attainable with top-down fabrication approaches which suffer from hurdles in multiplexing and hybrid material integration especially if sub-micro structures are needed. In addition, DNA self-assembly does not need any clean room facilities, thus justifying the lower costs associated with bottom-up approach in comparison with top-down.

The development of more in-depth applications in the bioanalytical field will require a number of issues to be addressed: cost of the DNA oligonucleotide synthesis, integration into devices and means of detection.

The first issue could be solved by employing cost-effective synthetic strategies such as DNA scaffolds synthesis by enzymatic reactions and construction of periodic nanostructures with reduced number of staple strands. The cost of staple strands synthesis can be reduced by employing direct printing techniques to dispense on a chip embossed with functionalized micropillars and amplified off the chip by nicking-strand displacement amplification.

In order to make DNA nanostructures usable in *Analytical Chemistry*, one has to envisage proper integration into analytical devices in order to extract signal output that can readable and easy to analyze. This is possible only if DNA nanostructures are accurately oriented on solid surfaces or solid-state nanopores in order to produce chips of DNA origami interfaced with electronic circuitry for signal extraction.

The last hurdle is the proper detection method. In the first approaches, researchers employed atomic force microscopy (AFM) (to measure physical properties like stiffness or hardness), and transmission electron microscopy as signal readout. Although AFM is employed as a readout in many reports, it could constitute a technical obstacle due to the slowness and difficulty in analysis. As an alternative to these methods, it has been shown in some very recent papers the possibility of performing detection (up to a single molecule) through DNA-based nanoplasmonics devices (Raman detection) or DNA origami single nanopores (electrical detection). DNA nanostructures have shown great potentiality as tools to enhance analytical performances (i.e. sensitivity and specificity) of standard bioassays with detection outputs such as quartz microbalance detection, voltammetry, colorimetry, etc.

DNA nanostructures offer one of the most flexible and adaptable approaches towards the assembly of tomorrow's new molecular machines and biosensors. The field is still in its infancy, especially as concerns the applications in *Analytical Chemistry* and thus, it also constitutes a new challenge for scientists who work in the field, given the necessity to evolve DNA nanotechnology from its status of beautiful and interesting subject to the one of practical and useful for Physicists, Chemists and Biologists.

Index

www.ingramcontent.com/pod-product-compliance
Lightning Source LLC
Chambersburg PA
CBHW050601190326
41458CB00007B/2136